Applied Literary Criticism

Joseph Chandra

ATLANTIC
PUBLISHERS & DISTRIBUTORS (P) LTD

Published by

ATLANTIC
PUBLISHERS & DISTRIBUTORS (P) LTD

7/22, Ansari Road, Darya Ganj, New Delhi-110002
Phones : +91-11-40775252, 23273880, 23275880, 23280451
Fax : +91-11-23285873
Web : www.atlanticbooks.com
E-mail : orders@atlanticbooks.com

Branch Office
5, Nallathambi Street, Wallajah Road, Chennai-600002
Phones : +91-44-64611085, 32413319
E-mail : chennai@atlanticbooks.com

ISBN 978-81-269-1855-3

Printed in India at Nice Printing Press, A-33/3A, Site-IV,
Industrial Area, Sahibabad, Ghaziabad, U.P.

Preface

This book has been fuelled by a passion for literature as it deals with the true interests of humanity, its recurring problems and the deep joys and fears of the human heart. Literature is an artistic depiction of human values. Without such values literature is reduced to journalism. Literature has a lasting value. Technology can become obsolete but not literature. For it is a contemplation of the nature and significance of our existence.

Literature is a source of valid knowledge as it deals with personal pain and happiness that are profound and universally felt. We feel the kindred throb of the human heart which, because human is neither Indian nor European, becomes a universal phenomenon. As the psychologist Carl Rogers once said: "What is most personal is most general".

The twenty-first century has been hailed as the age of Information Technology. In a highly automated world driven by microprocessors, literature still has its perennial relevance. The world has progressed tremendously in its outward material aspect, but there is no evidence of any great change in the inner consciousness. The primitive forces of hatred and envy still exist turning the earth into a valley of blood and tears. It is in the midst of such a sordid human condition that literature becomes a source of instruction on how to live life.

Matthew Arnold placed literature in a moral context when he stated: "More and more mankind will discover that we have to turn to literature to interpret life for us, to console us, to sustain us. Without poetry, our science will appear incomplete."

All the dry volumes of philosophy and ethics that were ever written cannot be compared to the intuitional knowledge of human nature we can acquire by reading *King Lear* and *Othello*. Shakespeare is more contemporary than Plato because each age finds something interesting and valuable in him. The French symbolist poet Paul Valery once noted that philosophers try to separate beauty from the things that are beautiful. This is philosophy's greatest drawback compared to literature as the things themselves become irrelevant. Literary texts, as this book demonstrates, provide a feast for the mind, heart and all the senses we bring to reading. Literature also enables readers to discern common problems, values, aspirations that all human beings share.

Literary criticism can be defined as the study, evaluation, and interpretation of literature. Modern literary criticism is often informed by literary theory, which is the philosophical discussion of its methods and goals. The two activities are closely related, but literary critics are not always, and have not always been, theorists. Some critics consider literary criticism a practical application of literary theory, because criticism always deals directly with particular literary works, while theory is often more general or abstract.

The book provides a refreshing introduction to literature and throws light on the application of theory to literary interpretation. The three genres of literature—poetry, drama and fiction—have been analyzed in the context of literary criticism. Some subversions of postmodern literature have been discussed and an exclusive chapter has been devoted to various aspects of short story. It will be useful for the students and teachers of English literature, particularly literary criticism and researchers in these fields.

This book makes no scholarly pretensions. It has been designed to serve as a sequel to my earlier book *Classical to Contemporary Literary Theory: A Demystified Approach*. The only aim of this work is to introduce literature as a unique discipline worthy of intellectual analysis as well as stimulating intuitive responses. It can be simultaneously reasoned out by the mind and keenly felt by the heart. Great care has been taken to present the subject in a lucid style to make it accessible to college students and general readers.

Joseph Chandra

Contents

Preface .... *iii*

1. Introduction to Literature 1

2. Applying Theory to Literary Interpretation 10

3. Reading and Responding to Poetry 28

4. Unlocking the World of Drama 45

5. The Linguistic Universe of Modernist Fiction 64

6. The Subversions of Postmodern Literature 75

7. The Long Delights of Short Story 90

References 95

1

Introduction to Literature

Literature is a fascinating discipline. It is primarily the communication of experience and not information. Understanding its nature and purpose helps us enjoy it fully. Rees defines literature as "the expression of thoughts and feelings of actual and possible experiences that are common to mankind". It does not concentrate on a narrow area of knowledge like physics or chemistry but with life in all its manifestations. It is concerned with all aspects of man and the universe in its entirety. Literature is a verbal art whereas painting is a visual art form. The Canadian critic Northrop Frye's statement that "Science can grow but literature is perennially alive" is a source of encouragement to all readers and students of literature.

It was Aristotle who first defined Art as an imitation and 'showing' of nature or "*mimesis*" in Greek. For Plato, Art was "*diegesis*". It was 'telling' about human actions or social life. It had to be reality-oriented. For Matthew Arnold, the Victorian poet and critic, "Literature is a criticism of life". The twentieth century moralistic critic Edmund Fuller defined literature as "a commentary on the life and nature of man". For F.R. Leavis, "Literature is a contemplation of life". The German philosopher Gadamer considered "Art and literature as a mode of experiencing reality". He considered literature to possess 'epistemic value' and valued it as a legitimate source of knowledge.

Martin Heidegger, the existentialist philosopher gave three lectures on *The Origin of the Work of Art.* He connects Art with truth. The essence of the artwork is not its 'representational' character but rather its capacity to allow the disclosure of a world. Truth here does not mean 'correctness' but 'unconcealment', "to make explicit" or "laying bare" what is concealed.

Heidegger in his essay *What Are Poets For?* cites Rainer Maria Rilke as an example of the highest form of thinker in his essay. Rilke is generally considered the German language's greatest twentieth-century poet. His haunting images tend to focus on the difficulty of communion with the ineffable in an age of disbelief, solitude and profound anxiety.

Literature generally deals with the five main categories in the whole spectrum of life.

1. Self—self-image, intra-personal reality, the inner world, etc.
2. Other—other people, inter-personal relationships, society, etc.
3. Life and Death—love and war, joys and sorrows, hopes and disappointments, etc.
4. World—The physical world, nature, etc.
5. God—Supreme Being, Being of Beings, Unmoved Mover, etc.

For Gadamer, experience of art is a matter of "ontological disclosure" or revelation of reality rather than a matter of subjective consciousness. He extolled the value of art by pointing out its capacity for "unveiling the truth" of human existence. The fact that truth achievable in no other way is experienced in an artwork constitutes the philosophical significance of art, which asserts itself against all rationalization.

In his *Map of Misreading* the renowned Yale critic Harold Bloom declared that, "Ordinary readers read while extraordinary readers misread". By misreading he did not mean getting facts wrong or losing one's perspective but active reading, that which brings the whole being of man into activity. A sensitive reader must be able to apprehend the value of a piece of literature. For a devoted reader of Shakespeare, there is something in Shakespeare's tragic heroes, which escapes analysis and labelling. A 're-cognition' takes place and one realizes that it is we who are sometimes Hamlet or Othello. Reading a good realistic novel not just cheers us up but rather strengthens our spirit, cleanses us of our negativity and elevates our Self.

Poetry, said Bloom, is the most supreme mode of transcendence. A good poem can possess us. The proper work of poetry is "to startle us out of our sleep-of-death into a more capacious sense of life". Great literature is always deeply anchored in social reality. We measure the value of poetry by the depths of its roots in reality. For Bloom, "Fantasy literature is a piece of nonsense literature". In an echo of Arnold, Bloom encourages us to cultivate an enjoyment of poetry. Arnold's "touchstones" provide us concrete examples of different kinds and degrees of literary merit. One can embark on a critical excursion into some of the most admired works of literature using this method.

In Religion and Literature, T.S. Eliot enunciated an interesting split in judgment. He wrote: "The 'greatness' of literature cannot be determined solely by literary standards, though we must remember that whether it is literature or not can be determined only by literary standards".

The moral approach to the reading of literature has the longest history of all the types of literary criticism. Plato was concerned with the moral effects of poetry in his *Republic*. Horace gave great weight to the usefulness and beauty of poetry. Dr. Johnson always judged the moral content of the writers he discussed in *The Lives of the Poets*. Arnold stressed on the importance of "high seriousness".

For the German philosopher, Jurgen Habermas, "Art is an illumination of life", it sheds light on social existence. It makes sense of human existence by shedding light on the human condition. It revitalizes our interpretations and perceptions. Thus, literature is not just an academic discipline. It has a practical, real-life orientation and prepares you for living the good life. William Walsh refers to the "culture of feelings" and the "training of emotions" frequently associated with literature.

Literature, which is concerned with all aspects of human living and the universe in its entirety, can only give us a complete notion of being human. It does not deal with "the modular man" (Alvin Toffler's term) but with the complete human being. In the words of Simone de Beauvoir, "Literature should deal with the most singular experiences to communicate the universal dimensions of the human condition". Another colossal figure among contemporary writers, Salman Rushdie defines literature as "a questioning of life". It asks fundamental questions about our existence. Great literature should move beyond the culture-specific conditions whether Indian or African and touch the heart of the human condition.

Orhan Pamuk, the Nobel Prize winner for literature in 2006 considers literature as "an enquiry into the soul of something as well as an exploration of the human condition." For him, "Literature is an honest attempt to unveil the truth". It is almost an echo of Gadamer's view. His novel *Snow* though a very political novel deals with the Appearance vs. Reality dichotomy. What appears bad can be really good and vice versa. When a woman character in *Snow* says to her lover, "My faith is more important to me than your love", the reader is not merely shocked emotionally but stimulated to interrogate or enquire into the very nature of faith and love. This novel has lines that you can carry in your heart for a lifetime according to a critic. This is literature at its highest philosophical level and affirms the German philosopher Schlegel's definition: "Literature is immortality in words".

For Jean Marie de Clezio, winner of the Nobel Prize for Literature in 2008, "Literature is primarily a translation of life into words."

The Nobel Laureate in Literature for 2010, Mario Vargas Lhosa says that the roots of literature may be in personal memory but it has a long range vision. He states that literature must transcend personality and reach the essential aspects of the human condition. He says, "Literature is a depiction of human existence."

The 2011 Nobel Laureate in Literature, the Swedish poet Thomas Transtromer defined literature as "an exploration of both Nature and human nature." Transtromer's poetry is universal and particular, complex and very direct at the same time. His works explore the themes of nature, isolation and identity. Given below is an excerpt from his poem entitled *National Insecurity*:

> The Under Secretary leans forward and draws an X
> And her ear-drops dangle like swords of Damocles.
> As a mottled butterfly is invisible against the ground
> So the demon merges with the opened newspaper.
> A helmet worn by no one has taken power.
> The mother-turtle flees flying under the water.

According to critics, his condensed translucent images give us fresh access to reality. His poems are characterized by very strong psychological insights into humanity.

The 2012 Nobel Prize for Literature was awarded to the Chinese novelist Mo Yan. He was praised "as an author who with hallucinatory realism merges folk tales, history and the contemporary." As a novelist he brought to life the turbulence of the 20th century China in vivid and often graphic works set against the tumult of the Japanese invasion and a struggling countryside. The hardships of life in rural China were captured in his first major work *Red Sorghum* (1987) which brought him nationwide acclaim. Among his more popular works is the *Republic of Wine* (1992) which lifts the veil on the drinking culture and the accompanying corruption that permeates all levels of government. In *Frog* (2009), he tells the story of a midwife haunted by the many forced abortions she witnessed. Thus his sweeping novels often reflected the turmoil of 20th century China from the cultural revolution of Mao Zedong (1966-76) to the horrors of family planning campaigns. His depictions are powerful and evocative which are the qualities of enduring literature.

Another important characteristic of literature according to the Russian Formalist critic, Viktor Skhlovsky is "*defamiliarization*". It is the capacity of art to counter the deadening effect of habit and convention by investing the familiar with strangeness and thereby deautomatizing perception. Defamiliarization is not simply a question of perception it is the essence of literariness. It is looking at the world anew—an attempt to discover the new world within the known.

When it comes to the actual teaching or learning of literature, both teachers and students have to be very clear about their objectives. The major objective in reading literature is not to coach learners for tests and examination but to enable them to derive a lasting satisfaction from literature. We cannot simply reduce literature to an examination system by converting it into facts. The Canadian critic Northrop Frye remarked that "Works of literature are not so much things to be studied as powers to be possessed". Another Canadian novelist and poet Margaret Atwood wrote: "A word after a word after a word after a word is power". Frye also claims that literature has a therapeutic function to perform.

The reading of literature can provide a continuous stimulus for thinking things over. We also develop notions of taste, the ability to assess validity and the recognition of beauty. Students are constantly prodded to read literature for marks; this has made the study of literature a very unpleasant exercise.

Existential Appropriation of a Text

Readers of literature must understand the fact that literature is no ordinary language but a special kind of language use with its own distinct linguistic organization and properties. What is crucial in the reading process is to be able to assemble and experience the meaning and not to simply abstract "the message". The radical shift must be from meaning extraction to meaning creation. As literature implies more than stating something plainly, meaning creation proceeds by relating one's understanding, knowledge and philosophy of life with the written text. I.A. Richards highlighted the value of reading literature very succinctly: "Literary texts render a number of discoveries for the reader, a relationship between book and the world where meaning is ever waiting to be formulated".

Too textual an approach to literature destroys its appeal to the general reader as well as students. Literature is Man's exploration of Man. Readers must get an experience of living through the novel or play and be able to identify with various characters in their conflicts and visualize life differently. We must constantly relate the text to life, to personal relationships, self-image, and a variety of social and moral issues, which make literature enlightening and enriching to us both emotionally and intellectually. After all literature is basically man's study of himself in an individual and social context. It is always within language that human experience is articulated. According to the philosopher and hermeneutician Paul Ricoeur, "The most important aspect of a literary text is its power to project a world as well as to serve as a mediation of self-understanding. To understand is to understand oneself before the text". He insists that whatever is intelligible is accessible to us in and through language and all developments of language like scripture and literature call for interpretation. "The essence of a literary work is to transcend its psycho-social conditions of production and be open to an unlimited series of readings" according to Ricoeur. Mere plot summaries are dead wood and readers must be initiated into the adventure of reading.

Wolfgang Iser, the Reader response theorist states that, "Reading is like a journey, reflections are like resting places in order to conceive ideas within". Thus, the issue of existential appropriation of a text becomes very crucial. Encountering a literary artifact should not be reduced to a mere intellectual or cerebral exercise. In the final analysis what counts is the degree and the manner in which the reader personalizes and assimilates the given text. Literature is a product of human experience and all human experience can only be communicated linguistically. Reading literature is a process of becoming human. English literature was introduced in British universities to replace classical studies. It took the place of the study of classical literature as a humanizing influence and to teach values.

Literature in the Twenty-First Century

The twenty-first century has been hailed as the age of Information Technology and Biotechnology. The Arts and Humanities can never be at the margins of university education in a technology driven and dominated world of our times. No society can develop ethically and culturally without the humanities. The promotion of values and ethics at different levels to make an impact on people's lifestyle and behaviour and help build a sustainable future is possible only through the study of humanities. The humanities can no more be considered a donkey on which the computer and IT rides but it should rather be the opposite.

The Nobel committee in Sweden still awards a Nobel Prize for Literature. Since 1901, 103 Nobel Prizes in Literature have been awarded. The two great physicists who won the Nobel Prize for physics in 2002 said in an interview that what made life meaningful for them was spending an evening listening to music or going for a walk in the countryside. One of them remarked, "Being a scientist does not disqualify you as a human being". The iconic inventor of the 21st century Steve Jobs declared while unveiling the ipad in 2010 that "Technology alone is not enough. It's technology married with liberal arts and humanities that yields the results to make our hearts sing."

At no time in human history has literature been challenged with a greater opportunity to uphold all moral and spiritual values, which make us really human. Technology is power but literature is the essence of culture and without its humanizing influences, science can only turn us into monsters. We cannot cut ourselves off from our own humanity and remake ourselves into efficient robots.

Contemporary Literature and Literary Theory

The study of literature is undergoing a crisis today. The great debate among critics is whether to treat literature as "pure" literature or merely as part of a larger framework called literary studies. As readers of literature we have to respond to this crisis. Literary studies have been literally hijacked by critics with political interests. The critics have grouped themselves under different labels and warring camps. One group is poststructuralist, another feminist or new historicist and so on.

Today we are not supposed to teach Shakespeare the old way. A Shakespeare play is seen not as a product of a single mind or a literary genius but as a product of social influences. We are compelled to examine the various socio-historical contexts that created a work like *The Tempest*. Therefore, instead of discussing the work as "pure" literature, we classify the work as postcolonial and use the theory as a frame of reference to read and explain the play. When we read Shakespeare's *Tempest* as a postcolonial text we may wonder whether we are missing the quintessential Shakespeare altogether. When we do not read Shakespeare as Shakespeare as Bradley read the four great tragedies of the Bard, are we then allowing literary theory to displace literature is a pertinent question.

When we begin to access literature through theoretical constructs we are faced with a dilemma. Literature departments have been renamed as Department of Culture Studies in the West. Literature thus becomes increasingly irrelevant to culture studies. In our post-postmodern world, the Harry Potter novel series and Batman comics may replace Shakespeare and Milton as the latest literary fad. Literature will survive but whether literary studies will survive is a major question. There is a danger of English Literature being taught by a specialist in future as Greek and Latin literature is taught today. Never before have we faced such an urgency to arrest the decline in the reading and enjoyment of great literature. We need to rekindle an authentic passion for reading immortal literary works from Homer to Hughes.

2

Applying Theory to Literary Interpretation

Reading literature as 'pure' literature is a very limiting experience in an era of theory explosion. There is a place for the intersection of theory with literature by relating it to any genre of literature—fiction, poetry or drama with a new and fresh perspective. Theory can in fact enrich our responses and evaluation of literature. Literature becomes intellectually challenging and theory becomes conceptually interesting when applied to the interpretation of a literary text. The possibilities are endless with the application of theory to literature and the reader is introduced to the literary delights of new perceptions and fresh insights into life and the universe. The German philosopher Gadamer extolled the virtues of Art and literature by stating that "Works of art carry with them an invitation to be interpreted, not in an 'objective' way but as 'a conversation' between those whom it touches. This is the 'true' meaning of art and by extension of life. Philosophy is speculative, experience of literature involves engagement".

It must be borne in mind that no school of literary theory is the "final" perspective or approach to a literary work and no single critic has all the answers or can provide a definitive analysis of any text. Literary criticism today has become "The Library of Babel" (Jorge Luis Borges' term) where each critic and school of literary thought cancels out the value of each other. Formalism and Reader-response criticism are at odds with each other just as poststructuralism subverts the major premises of structuralism.

Of course, there is some validity in what each of these schools of thought theorise about literature. It is hard to be totally wrong. But in the face of so much opposing claims the reader is bewildered and thrown into a state of confusion. We live in a climate of ideas where one idea is pitted against another. Literary criticism is an inquiry into the fundamentals of any X. A reader of literature is sensitive to the language of the text. There are many reading options open to him. No school of theory should be taken for granted. The need of the hour is conceptual eclecticism. It is an eclectic approach based on micro-components in contradictory systems. This is the best way to handle a crisis engendered by different possibilities of reading a literary text.

The deconstructionist critic Paul de Man lets his own readings of literature to be deconstructed by other readers and critics. He considers the first reading of a text as an elementary reading. It serves only a utilitarian purpose for our students by offering a paraphrasable summary for obtaining marks in an examination. The second reading is a meta-reading of the text. It makes us intellectually vibrant and helps us explore new dimensions of the text.

A brief introduction to the different schools of literary theory and how applying them to a text can transform the reading of literature into a source of edification and enjoyment will be demonstrated in a simple and direct manner. An elaborate study of these schools of theory can be found in my book entitled *Classical to Contemporary Theory: A Demystified Approach* co-authored with K.S. Antonysamy. One should not study theory for theory's sake but use theory as a framework of reference for reading literature.

Postmodern Sensibility and Responses

The great critic and novelist Umberto Eco makes a crucial distinction between two kinds of literary texts—an *"Open Work"* and a *"Closed Work"*. In an "open work" there are two kinds of freedom:

1. Freedom to interpret
2. Freedom to have one's own judgement

He cautions us that it should be a studied endeavour and not intellectual licentiousness. This results in an act of improvised creation based on the reader's expectations—the politics of the reader operating all the time. Eco declares that, "A work of art is a complete and closed form in its uniqueness. It is an organic whole". An "open work" on account of its countless interpretations leaves the "right" interpretation out of the question.

Eco asserts that all our readings are products of our own ideology, social and class influences. For instance, after reading Shakespeare's *Antony and Cleopatra,* a conservative Christian might conclude that Mark Antony is an immoral man and his adulterous affair with Cleopatra was the cause of his downfall. A reader or critic with liberal views on the other hand might question the monolithic conception of tragedy expounded by Aristotle and prefer the Hegelian concept of tragedy instead.

In classical Greek tragedy, the tragic hero (a gender-neutral word) was the victim of Fate or external circumstances beyond his control. In Sophocles' *Oedipus Rex*, Fate overwhelms Oedipus and kills him finally. The protagonist is always a lonely figure pitted against the cosmic powers in a losing battle.

A.C. Bradley while commenting on the role of Fate in Shakespeare's tragedies remarked that had Desdemona not dropped her handkerchief in the most vital spot she would have been alive. In the case of Romeo and Juliet, Fate basically played the role of "Bridge to Heaven". The love between Romeo and Juliet was so pure that they deserved heaven and the sin-stained earth was not a fit place for them. Bradley had an extensive influence on twentieth-century criticism. Bradley's *Shakespearean Tragedy* has been described by Terence Hawkes as "one of the most influential texts of our century". Though his focus on character has been extensively challenged, Harold Bloom calls Shakespeare as "the original psychologist". Jonathan Dollimore calls Bradley's character criticism as the 'metaphysic' of tragedy.

Shakespeare's plays can be considered as open works. Harold Bloom in *Shakespeare: The Invention of the Human* argues that "Whether we are male or female, old or young, Falstaff and Hamlet speak most urgently for us and to us. Falstaff, at his funniest or at his most reflective, retains a vitalism that renders him alive beyond belief. When we are wholly human, and know ourselves, we become most like either Hamlet or Falstaff". Bloom concludes that the most 'round' male characters in Shakespeare are Hamlet and Falstaff and Cleopatra dominates the women.

For Aristotle, the tragic hero was always a victim of *'hamartia'* or a tragic flaw to which he is blind. Fate thus becomes an internal dynamic or a character crisis that when unrecognized and uncorrected destroys the hero leading to a *'catastrophe'*. But for the German philosopher Hegel, tragedy is not a conflict between good and evil but a collision between equal and contrary values. The issue in tragedy is a struggle between the right and the right or between the right and the good. The hero is caught in "a collision of equally justifiable moral aims". Thus a liberal critic may raise a pertinent question, "Is Antony immoral?" He may even consider Cleopatra as a fever from which he could never recover. The play could be interpreted as a conflict between a lesser good, his duty to Rome and a higher good his search for true love and fulfillment in life. A moralist reading of the play would be just the opposite. The play will be read as a conflict between duty and beauty or responsibility and pleasure. All interpretations lead to an indefiniteness.

According to Eco, the Symbolists were the first proponents of an open work. Mallarme insisted that the primary task of a Symbolist poet was "to suggest the pleasure of guessing, never to name, always to suggest, for plain statements do not make great poetry". A work of art can be interpreted in any way you choose.

In an *"open work"* you always get a fresh perspective of life. There are no canonized responses. The first sign of postmodern sensibility is the encouragement of acts of conscious freedom and 'hijacking other discourses'. There are questions of valid interpretations involved here.

A *"closed work"* on the other hand is a finite work. Eco describes a closed work as "fixing limits of discourse". An *"open work"* allows the endless construction of new discourses. It is a work in movement, with no inertia. It contains within itself, "the infinite treasure of buried significations" (Foucault's term).

Roland Barthes made a distinction between *"Readerly Texts"* and *"Writerly Texts"*. Readerly texts impose a meaning on the reader by encouraging passive consumption. They were designed to constrain the exercise of the reader's imagination. Writerly texts invite the reader to engage in the production of textual meaning. A *"closed work"* dissolves into an *"open work"* for both Barthes and Derrida. Barthes wrote that "If words only had one meaning (dictionary meaning) there would be no literature. Literature is based on the very plurality of meanings. No one can arrest the flow of meanings in a text". The etymology of the word text is from the Latin *'texere'* which means to weave.

For Eco, a work is eternal not because it imposes one meaning only but because it suggests different meanings to different readers. He relates every work of art published to the latest scientific findings in physics, astronomy as well as from a political and philosophical perspective. James Joyce's *Ulysses* lends itself to multiple interpretations when read as an open work. Reading thus becomes a difficult and a disciplined activity.

In Shakespeare's *Macbeth* the leit motif of the play is the Appearance vs. Reality dichotomy—what seems as real is not the truth. For Eco, the indeterminacy in the reading of a work of art lies in the multi-value logic one adopts. An open work refuses to implicate itself to any one ideology. Reading literature thus becomes an inter-disciplinary experience. Samuel Beckett's *Waiting for Godot* is an open work that cannot be completed or be completely read. No reader can pin down any definite meanings. It has a 'playful pluralism'. It allows an immense amount of freedom to the reader. Paul

Ricoeur calls this phenomenon, "the semantic plenitude of the text". The text is open to an unlimited series of readings.

A closed work is so well structured verbally that it is not dynamic enough for multiple readings. It cannot be open unless it is rewritten but do we have the freedom to alter the work beyond recognition is a matter of debate.

Formalism

Formalism insists on a strict "word-on-the page" approach. It stresses on the importance of literary form in any discussion of meaning. Biographical, historical and socio-cultural aspects are considered to be irrelevant details in determining the literary merits of a work. It made a distinction between literary writings and sub-literary or non-literary writings. Formalist critics privileged poetic language over all other forms of discourse. The use of language in a telephone directory or an operations manual of a car has no aesthetic pretensions. It is entirely functional language intended to communicate information.

A poem is considered a verbal icon or a carefully constructed piece of verbal art. Literary language is a kind of "linguistic sculpture" to be investigated very seriously through the technique of "close reading". The Russian Formalist Roman Jakobson defined poetry as "language in its aesthetic function". One of the principal aims of Formalism is the scientific study of literature. This objective examination of literature would enhance the reader's capability to read literary texts in an appropriate way with an eye for those properties of the text that are considered 'literary or artistic'.

For example, a formalist while reading one of Shakespeare's sonnets that describe the Dark Lady would consider as irrelevant whether she was a historical figure and dismiss questions of personal identity such as "Was she a Black woman called Lucy Negro?" Instead a formalist critic focuses on reading the text closely, pays more attention to linguistic organization and nuances of meanings generated by word choice and the writer's use of imagery, paradox and irony. The American formalist movement was called New Criticism.

Today reading literature goes beyond a strictly formalist approach. Reading literature without taking into account the historical and cultural reality behind the production of a literary work (its relation to extra-textual reality) is only empty formalism, which fails to enrich our understanding of literature. Without Socio-historic contextualization in New Criticism an alternative dimension remains unexplored.

Marxist Criticism

Marxist critics usually apply the social and economic theories of Karl Marx and Friedrich Engels to their readings of literature. Marxist literary theories usually focus on the representation of class struggle in their readings of poetry, fiction and drama. Any great work of literature can be studied for their treatment of class conflict and other Marxist concerns.

Marx believed that the dominant capitalist upper class would be challenged and overthrown by the working class. The working class who were exploited by the capitalists would eventually create a classless society. Marx highlighted the alienation of the working class who in spite of producing all the products and profits did not share in the benefits of their labour. Raymond Williams and Terry Eagleton are the well-known Marxist critics in the United Kingdom and Frederick Jameson is the best known American Marxist critic.

A Marxist reading of Charles Dickens' *David Copperfield* might concentrate on how David Copperfield as a young boy was directly affected and exploited at the blacking factory by the English capitalists. The grim reality of hand-to-mouth factory existence of David shows the horrible conditions of the working class during Victorian times. A Marxist critic would point out the clear divisions between the capitalists and the toiling classes worsened by the pressures of the economic system of which the latter were victims.

According to the French philosopher Paul Ricoeur, Marxist criticism is based on the *"hermeneutics of suspicion"*. This approach concerns itself not only with what the text says but with what it hides. It differentiates between the "Overt" and the "Covert". It focuses on literature as a cultural production revealing a form of ideology, values, feelings, etc. It examines how class conflicts, tensions and contradictions between the working class and the capitalists are represented. It uses traditional techniques of literary analysis to highlight social and political meanings. The Hungarian Marxist critic George Lukacs remarked that Marxist criticism cannot function outside of the strict Base/Superstructure model. Base is the economic system on which superstructure or culture rests. For Antonio Gramsci, the Italian Marxist theorist, Hegemony is a sort of deception in which the oppressed forget their own conditions and accept the dominant values of the oppressor as their own. Marxist literary criticism has influenced feminism, new historicism and cultural studies. Thus Marxist literary criticism focuses on society as it is reflected in literature.

Psychoanalytic Criticism

Psychoanalytic criticism analyses literature as an imaginative expression of the inner workings of the human mind. It is strongly influenced by the ideas of Sigmund Freud. The basic premise of Freudian psychoanalysis is the tremendous power the unconscious mind has on our conscious life. The conscious mind is only the tip of the iceberg. Our desires which are socially unacceptable are repressed and pushed into the Unconscious. In Freud's tripartite structure of the human psyche; the Ego which is the conscious part of our mind serves as a mediator between the Id which consists of our biological drives and the Superego which plays the role of the censor or conscience by forcing much of our desires into the unconscious for the sake of social harmony. The superego represents the prohibitions of society, religion, etc. Freud theorized that such repressed experiences surfaced through dreams. Psychoanalytic critics therefore interpret literature as the reflection of our unconscious life.

Freud used the techniques he had developed for interpreting dreams to interpret literature. He also wrote critical commentaries on Shakespeare's great tragedies, *Macbeth, King Lear* and *Hamlet*. His insightful remarks on

Dostovoesky's *The Brothers Karamazov* are a brilliant example of psychoanalytic criticism.

A psychoanalytic study of William Golding's *Lord of the Flies* would explain the three major characters Ralph, Jack and Piggy as symbols of the Ego, Id and Superego. When Jack takes over the island after killing Piggy and Simon, it represents the total dominance of our biological drives over the intellect. The island goes up in flames and Ralph has to run for cover to save his life. According to the psychologist, Richard Hughes, "The children in the novel are not a bunch of small, fanciful adults but a miniature version of a savage race". According to Freud, the dominant part of Man's psyche is irrationality and not rationality. Civilization is based on repression and if allowed uncontrolled expression will only result in the crime graph shooting up with more incests, rapes and murder.

Structuralist Literary Criticism

Structuralism is a school of thought with roots in linguistics and anthropology. It is a theory of language that defines language as a system of signs that have no inherent meaning in them except in "an agreed upon or conventional relation to one another". The linguistic model is used as an analogy to study literary texts. Structuralist critics make no claims about possessing a new method to interpret literary works but rather demonstrate how works of literature transmit meaning.

Structuralist literary critics view poems, novels and plays as part of a larger literary system and not as individual self-contained works with inherent meaning. No reader can analyze and appreciate a literary work without understanding the system within which it operates. Both linguistic and literary structuralism stress on the importance of difference. Without understanding the difference between the structure of poetry and prose one cannot make any sense of lines like this:

> Axes
> after whose strokes the wood rings
> and the echoes!

Echoes traveling
off from the center like horses. (*Sylvia Plath, "Words"*)

Only those readers acquainted with the conventions of poetry will find these lines meaningful. We can interpret this group of "signs" only if we understand the system that governs the structure. Structuralist literary critics make a distinction between everyday language and literature. Though both use the same "signs" the purpose of ordinary language is to give information and literary language is intended to create aesthetic effects. The first stanza from Sylvia Plath's poems *Words* can be understood only by our recognition of the poetic form, the rhyming conventions and the figures of speech the poet uses. Thus a structuralist reading tries to view literature as a meaning system and objectively searches for principles of order, coherence and meaning overlooking barriers of time and culture.

Poststructuralist Reading: Deconstruction

Poststructuralism departs from the foundations of structuralism. Saussure considered the language system as a closed, stable system with a delimited structure of meaning. Language functions by the arbitrary connection of words (signifiers) to ideas (signifieds). Deconstruction, which is the Derridean school of thought, is one branch of post structuralism. It ruptures the bond between the signifier and the signified by arguing that as the connection between the word to an idea is purely arbitrary, any operation of language is inherently unstable. Deconstructionists insist that to analyze a literary text is to be confronted with a play of meaning or a continuously shifting set of meanings.

Deconstructionists believe that any given text has the potential to offer divergent readings, all of which are equally valid yet may in some way undermine and oppose one another. Deconstruction is not a school of criticism but a way of reading that takes apart a literary text and reveals its separate layers.

When a literary text is deconstructed it is being untied. The meanings that you find are different from the author's original intentions or meanings. Hillis Miller, a deconstructionist critic stated: "I do not dismiss intention but it can no longer dominate the site of articulation". What he meant was that the author cannot preserve his own meanings from the active interpretation of the reader.

Deconstructive critics refuse to accept a fixed meaning or any one way of reading a literary text as absolute. Their major contribution is the insistence and demonstration of the possibilities for multiple meanings within texts. Any text can be deconstructed. Deconstruction is a strategy of critical analysis. It is not a method and cannot be transformed into one. To deconstruct is to undo and not to destroy. It is destabilizing meanings constantly. For example, a reading of the first chapter of Genesis from the Bible will expose the hierarchies of binary oppositions that dominate the philosophical tradition of our thinking. We can challenge them but to destroy them is not possible without becoming psychotic.

In the beginning God created the heavens and the earth. And the earth was without form and void.... God said, 'Let there be light and there was light; And God saw the light, that it was good, and God divided the light from darkness...the greater light to rule the day and the lesser light to rule the night.

And out of the ground the Lord God formed every beast of the field, and every fowl of the air; and brought them unto Adam to see what he would call them; and whatsoever Adam called every living creature that was the name thereof.

And the Lord God caused a deep sleep to fall upon Adam and he slept; and he took one of his ribs and closed up the flesh instead thereof. And Adam said, "This is now bone of my bones and flesh of my flesh; She shall be called woman because she was taken out of man. And Adam called his wife Eve" (*Genesis, Chapter – 1, Holy Bible*).

The following binary oppositions are identified, carefully examined and the privileging of one over the other is critically put under interrogation.

Heaven / Earth Form / Void
Light / Darkness Man / Woman

Derrida in his book *Of Grammatology* cautioned readers about totally abandoning these divisions in our thinking. He wrote: "It is not a question of rejecting these notions, they are necessary and, at least at present nothing is conceivable for us without them.... These concepts are indispensable for unsettling the heritage to which they belong" (*Of Grammatology*, p. 14).

After Derrida, the most well known deconstructive critic is Paul de Man. He demonstrates how the two modes of language, "literal" and "figural" can give rise to diametrically opposite meanings in a 'text' or 'verbal event'. Derrida on his part shows how language as a 'system of infinite difference' renders meaning problematic. For Paul de Man a deconstructive reading often turns on the literal/figural distinction. This could result in an *'aporia'* or a figuration of doubt that allows for two incompatible and irreconcilable points of view.

Paul de Man offers an illustration of how deconstruction works by attempting an interpretation of the last four lines from Yeats' poem *Among School Children*:

> O Chestnut tree, great-rooted blossomer,
> Are you the leaf, the blossom or bole?
> O body swayed to music, O brightening glance,
> How can we know the dancer from the dance?

The usually accepted reading of this poem, as Paul de Man points out, considers these questions to be rhetorical (i.e. figural), as an assertion, on the one hand, of the potential unity between form and experience—*'structure'* and *'event'*, in Derridean terms or between *'sign'* and *'referent'*. But on a literal reading, again as de Man shows, the lines can be treated as genuine questions, raised in desperation and as such they acquire a ring of urgency.

The first of the two questions can be answered by explaining the allusion to a Greek legend about Attis, the god of vegetation, holding as he does the fiery blossom and earthly leaf at once together and apart. The second question reveals the deconstructive reader's predicament. How can we possibly distinguish between the literal and the figural and between 'sign' and 'referent'? A literal reading of the poem is shown to reverse the 'mood' and 'mode' of the poem, a reversal not certainly intended by Yeats! De Man cautions us that neither reading (literal/figural) is privileged. Hence, the 'aporia' or uncertainty faced involving doubt is encountered. The deconstructive reader would not be able to decide between the two readings. He would remain in an epistemological dilemma forever. Thus a deconstructive reading shows how language may and does generate meanings, even diametrically opposite ones on its own though the intention of the speaker/writer may be different.

Postcolonial Criticism

Postcolonial literary criticism is a type of cultural criticism which involves the analysis of literary texts produced in countries and cultures, which had come under the control of European colonial powers at some point in their history. Literally, Postcolonialism refers to the period following the decline of colonialism or the end of domination by European empires.

Postcolonial literary criticism has been influenced by Marxist thought, by the work of Michel Foucault and by deconstruction, which has challenged not only hierarchical binary oppositions such as West/East and Centre/Periphery but also the notions of superiority, associated with the first term of each opposition.

A postcolonial reading of Arundhati Roy's *The God of Small Things* would interpret the novel as a kind of Protest or Resistance Literature. It is a 'literature of rupture' entirely different from the Romantic 'literature of rapture'. It is an Activist Literature that struggles for social justice. It is a struggle for a just social order that never ends.

Roy's novel cannot be read without empathy. It is the voice of the voiceless and articulates the experience of deprivation and dispossession. It is sub-altern literature that looks at human sufferings beyond socio-cultural barriers. The right to be human is at the centre of all postcolonial literature. A postcolonial reading accepts states of marginality, plurality and perceived 'otherness' as sources of energy and potential change.

The God of Small Things is all about sub-alterns, children, youth, women and untouchables. The narration is morally strenuous and there is an implicit tragedy even in the comical episodes. It cuts through the trappings of caste, class and religion and reveals the bare bones of humanity. Velutha, a Dalit is the protagonist of the novel. He is denied a social space by the so-called "Touchable kingdom" of upper caste folks, politicians and policemen. The mutual love and affection that Velutha and Ammini, is described as the solidarity of the dispossessed. Ammini is a victim of an abusive marriage. Velutha cares for her children also and plays with them regularly as a fathersubstitute. He is also a communist party cardholder. When Velutha is arrested on false charges and sent to jail, his party leader Comrade Pillai takes no steps to release him. Roy sarcastically writes that, "Another religion turned against itself. Another edifice constructed by the human mind decimated by human nature".

The police beat him up and torture him. The word 'dalit' itself means broken people. The police are described as "history's henchmen" attempting to instill order into a world gone wrong by humiliating and terrorizing Velutha. It was "Power's fear of Powerlessness". The heart of darkness was at work. The oppressor creates myths to perpetually enslave the oppressed who simply swallowed them without reasoning or questioning them. Velutha dared to be different from the herd. This was seen as a serious crime. In Velutha's death, "the Touchable Kingdom" had given a history lesson for all future offenders.

The novel abounds in accurate descriptions of Nature. Kerala becomes a microcosm of India. It is a masterpiece, a sad story told very tenderly by inventing its own language. Resistance Literature always creates a counter-discourse. Thus the oppressors' ideologies to subjugate the oppressed inflicts deep damage on the collective psyche of the depressed classes.

The only hope lies in refusing to accept one's condition of permanent dispossession and pain as a way of life.

Ecocritical Reading of Literature

Ecocriticism is the study of literature from an ecological or environmental perspective. In an age of environmental destruction, ecocriticism stresses on the necessity for a humane understanding of our relationship with the non-human world of nature. The term "ecocriticism" was first used by William Rueckert in 1978. In 1989, Cheryll Glotfelty in the *Ecocriticism Reader* defined ecocriticism as "the study of the relationship between literature and the physical environment". Reading literature from an ecocritical perspective is a highly enjoyable and enriching experience. "Literary ecology" employs ecological principles to read literature.

Ecocriticism is not just analyzing nature in literature but it implies a move towards a biocentric world-view as opposed to an anthropocentric view and calls for a broadening of human kind's conception of a global community to include non-human life forms and the physical environment. This is a matter of ethical commitment and calls for an extension of ethics that re-explores the environmental concerns and challenges the hegemony of anthropocentric constructions of the physical world.

Ecocriticism is not a recreation of Romanticism. Romanticism had different roots. Wordsworth's poems and the writings of Emerson as well as Thoreau on the relationship between man and his surroundings (Nature) are very different in scope from the ecoliterature of our times. Their reflections emphasize the mystical ties between Nature and human nature. They sought to renew the self by a "return to nature".

Contemporary writings such as Brian Clarke's *The Stream*, Jonathan Bates's *The Song of Earth*, Thom Hartmann's *The Last Hours of Ancient Sunlight,* Richard Heinberg's *The Party's Over* and Don DeLillo's *White Noise* and *Under World* with their ecological themes and ecophilosophical musings have created a new metaphysic in literature that has been described as biocentric writing. These writers are called ecowarriors whose literary activism is meant to draw attention to the rapacious economic exploitation of nature with scant regard for the environment or the future of humankind on the planet. These writers have produced critical literature which are the "landmarks" of ecocritical thinking. Such distinct 'greening' of literature energises our engagements with the world of Nature.

Don DeLillo is a post-war American novelist who presents a deeply disturbing apocalyptic novel *White Noise* (1985), highlighting the man-made catastrophe due to a lethal chemical leakage which is the inevitable byproduct of dizzying advances in technology. Though humans have taken rapid strides from the Stone Age to the Post-Industrial Age, it is a grim fact that we have traded the pastoral, idyllic, uncontaminated world of the past for a world, which is industrial, technological and pervasively toxic and hazardous. Gaining momentum from the industrial revolution, man has constructed technological marvels of extreme dimensions, large dams, superhighways, tall skyscrapers, and the frightening aspects of nuclear technology and genetic engineering. But we hardly realize that some of this dehumanizing technology with its rampant pollution and the consequent environmental catastrophe has begun to boomerang on the very roots of our existence. We are precariously poised between progress and the final nightmare hurling us into a vortex of forces leading to total annihilation. The industrial toxins which is a symbol of the evil consequence of mindless technological development in the novel develops a sinister and ominous edge inviting us to reflect on the importance of preserving our ecosystem from life-threatening pollution, if we are to survive on this fragile planet.

A "green reading" of Don DeLillo's another novel *Underworld* highlights the importance of sustaining a healthy web of life on Earth and warns of the total consequences of our callous waste disposal habits. DeLillo takes a radical break from the pastoral and ectopian rendering of nature in English Romanticism. William Wordsworth no doubt set forth a tradition of environmental consciousness and fought a single-handed crusade against the brutal attack on nature in the wake of the Industrial Revolution. His poetry, letters, pamphlets and other prose works created sensitivity to ecological issues in his period.

The contemporary American poet Richard Wilbur like Wordsworth draws almost all his imagery from nature and calls for a paradigm shift from ego-centered humanism to eco-centered humanism. His nature poems in a postmodernist context provide a fascinating glimpse of the eternal through the temporal.

Wordsworth is considered as "the first exemplar of ecocritical writing or Oikopoetics with his concern for the preservation of traditional rural ways of life". English Romanticism inaugurated a radically new conception of human kind's relationships to the natural world. Oikopoetics or ecopoetics from the Greek word '*Oikos*' which means home or hearth stresses on an environmental approach to literature. Don DeLillo's environmental imagination has an eschatological vision of planet earth. His writings can be compared to Blake's visionary protests against the Industrial Revolution.

Environmental literature problematizes the absence of land health and the despoiling of natural habitats. The novel *Underworld* in which garbage, refuse, nuclear waste, debris (of all kinds), dumpsites and landfills 'litter' the pages presents a mountain of ecological anxieties. Nick Shay, a waste management specialist, Arthur Blessing, the CEO of a waste management company and Jesse Detwiler, a self-styled garbage archaeologist offer nuggets of philosophy on living with waste and waste disposal. DeLillo reveals how urbanization and dependence on technology for increasing human comforts have led to the degradation of our Oikos. Instead of pontificating, he presents waste as a dominant metaphor of our Digital Age. This novel attempts to raise ecological consciousness with the hope of generating an ecological conscience. It stresses on the importance of a land ethic and waste consciousness to preserve our biosphere. It valorizes transhuman magnificence as opposed to technological dominance and makes it an integral part of the environmental discourse.

Environment figures in literature in several ways as theme, setting, persona (animal, bird, plant, landscape, season, cloud, sky, sea, mind, etc.) symbol/image, genre (pastoral, almanac), and also as device (pathetic fallacy). Throughout the ages, environment has figured in literature as theme. The most common ones are: nature versus culture, wilderness versus garden, fertility versus sterility, etc. However, in most recent times literature reflects the environment crisis thematically. Accordingly, degradation as in *The Waste Land* or in John Betjeman's poems and pollution as in Don DeLillo's novels and other issues such as loss of habitat in Amitav Ghosh's *The Hungry Tide* (2004) and regeneration figure in contemporary literature. The solution offered is to incorporate the dimension of the spirit into ecology to find a perfect balance between sustainable development and preservation of the environment. This new conception called "*deep ecology*" (Arne Naess' term) is the major focus of "environmental literary criticism" or "green cultural studies". Thus ecocriticism has altered our perception of literary studies.

Along with these schools of literary theory other approaches like New Historicism, Cultural Materialism and Feminism can be adopted to widen our mental horizons and further enrich our enjoyment of literature.

3

Reading and Responding to Poetry

Poetry is an application of the literary maxim: "Less is more". It is the most compressed form of literature. No definition can precisely define the nature and function of poetry. It is the oldest of all literary genres. Right from Homer onwards, poets have been image makers possessed by the power of language. Viktor Skhlovsky, a Russian formalist critic pointed out the fundamental difference between 'poetical language' and 'practical language'. Poetical language is 'non-casual and purposeful' whereas practical language is a casual verbal structure employed in the language of ordinary, everyday speech. It is not primary language that communicates information but is "language plus" that produces evocation.

F.R. Leavis insisted that any discussion of poetry should connect three things: The reader, the poem and what the poem is about. Reading and responding to poetry is 'a common pursuit'. The meaning of a work of verbal art like a poem is realized not in isolation; as something locked up in an individual's mind but as identification with the shared experiences of humanity. Carl Rogers' remark on counselling, "What is most personal is most general" is also true in the case of poetry. Poetry establishes a connection of minds. Poetry does not offer any conceptual knowledge but existential or experiential knowledge. It is not approximate knowledge or knowledge about things but proximate knowledge or knowledge of things. Meaning is provided not as explanation but as an experience. According to Paul Ricoeur, "It is only within language that human experience is articulated".

For the French postmodern philosopher Francois Lyotard, "Art is an explanation of life" and literature is a 'verbal event'. By 'event' he meant something happens that has the potential to shatter prior ways of explaining and making sense of the world. It calls for new modes of experiencing. Art can evoke feelings of disturbance and disorientation that are irreducible to rational thought or calculation.

Works of art and literature do not offer direct answers to political and philosophical problems. Rather, their value lies in an ability to generate questions that can challenge ways of thinking and genres of discourse that claim to provide total answers and systems. Poetry is a linguistic adventure but a painful quest for the right words.

English Poetry of the Twentieth Century

Modern poets turned more inward as modern sensibility became more complicated. Romanticism started with Gray and ended with Tennyson. The language of the Romantics had lost its force and was not adequate to express the complex sensibility of our modern times. Modern poets became fed up with the Romantic style of writing and the depiction of outer life. Romanticism had exhausted itself and modern poets began experimenting with language and to replenish it to express the modern sensibility. The characteristic twentieth century refusal of romanticism finds expression in the poetry of Eliot, Larkin and R.S. Thomas.

T.S. Eliot and W.B. Yeats were influenced by the Symbolist Movement in France. Arthur Symmon's book *Symbolism in Literature* was dedicated to W.B. Yeats who was called the first Modernist in English poetry. French symbolist poets like Baudelaire, Mallarme, Valery and Verlaine produced poetry that was very different from the poetry of the previous centuries. A poem need not be expressive but be evocative and suggestive. Baudelaire's "*System of Correspondences*" explained how the poet borrows images, objects and materials from the natural world to serve as a means of expressing his soul. Images and symbols are interchangeable terms. When an image persistently recurs in a literary work, it becomes a symbol. According to Ezra Pound, the founder of the Imagist movement an image with its visual components appeals to our sensory imagination and is an association of ideas and feelings. Images are used to create word pictures which will communicate some idea or attitude. Words which make a direct appeal to any of the five senses are images. Most images are visual. Images are created by the use of common nouns since they name concrete objects. Abstract nouns like 'honour' or 'courage' do not provide images. A symbol refers to something other than it is. A rose refers to beauty and a dove is an emblem of peace. The Cross stands for Christ and the Christian faith. Such a "concrete construct" in the form of highly charged symbols and images alone could capture the inner-stuff of the human psyche.

Yeats defined a poem as "a complex arrangement of symbols". Yeats gave the symbolist movement a spiritual touch. Symbols were considered as pre-logical devices capable of arresting the inner thought processes. Deep human feelings cannot be accurately expressed through direct language but can only be suggested through symbols.

No symbol had a fixed or a straightforward meaning. They had multiple layers of meanings like the smile of Mona Lisa. A poem thus becomes a water-balloon depending on the way you handle it. Both his poems *Byzantium* and *Sailing to Byzantium* are a string of symbols with no visible connection. In *Byzantium* he moves steadily through a series of symbols to communicate what he intends. As he surveys the place, he sees a mummified body which becomes a symbol of the meeting point of Death and Life. It is a death-in-life and life-in-death kind of existence. The perfect shape of the dome of St. Sophia Hagia's cathedral is a symbol of Art. The dome which is unchanging disdains the moon which is a symbol of flux in life. It waxes and wanes as it moves around the earth whereas the dome like the golden birds in the carved petals symbolizes the perfection and immutability of Art.

Common bird or petal
And all complexities of mire and blood
The dolphin-torn and gong-tormented sea,

The "golden birds" are compared to common birds and transient biological life with the immortal life of Art. Mortal life is frustrating and decaying. Every symbol in the poem reminds us of the transience of life in this world. Human life is symbolized as a perpetual struggle between sensual desires symbolized by the exuberant dance of the dolphin and spiritual yearnings symbolized by the gong. Yeats is trying to express through language a reality that transcends language. It relies heavily on techniques of definition by negation. Yeat's best poems defy analysis as shown in the lines below:

Before me floats an image, man or shade,
Shade more than a man, more image than a shade.

T.S. Eliot uses the same technique to survey the panorama of modern life in *The Wasteland.* The heap of broken images in Eliot's Wasteland symbolizes the discarding of traditional moral and spiritual values held dear by our ancestors. There is an obvious infertility in the carefully chosen symbols of *The Wasteland.* The continuous planting of corpses with no resurrection like the resurrection of Christ symbolizes the all pervasive spiritual barrenness of our times which produces only fear, death and damnation. The Game of Chess in the poem with its literary and cultural associations symbolizes the state of degenerate humanity with all its sexual perversions captured in the phrase "women beware women".

In the *Fire Sermon,* the imagery of young boys and girls meeting at the banks of the Thames as a rendezvous zone, the carelessly thrown cigarette butts and rusty juice cans suggest not only the physical filth accumulated in modern cities but also the moral decadence that accompanies it. Sex has become a casual and mechanical activity devoid of emotions and commitment. The Red Rock symbolizes Christ and water symbolizes regeneration through faith in Him.

The Waste Land expresses the disillusionment, sterility and triviality of contemporary life. It is a poem about two quests. First, the spiritual quest of the soul for salvation, and second, the intellectual quest of the mind, towards comprehension and articulation. It describes modern man as alienated from God. The poet laments on the exclusion of the sacred in the modern consciousness.

Eliot through his intellectual symbols succeeds in expressing complex philosophical and theological ideas. The thinking mind and the feeling heart were in dissonance with each other in both Neo-Classical literature (with its predominance of thought) and in Romantic literature (with its excessive emphasis on feelings). This *"disassociation of sensibility"* was rectified with a *"unified sensibility"* found in the seventeenth-century Metaphysical poets like Donne and Marvel and later in the symbolist poetry of Eliot and Yeats.

In the poetry of Eliot and Yeats one finds the thinking, word-arranging and clarifying power of a well-crafted work. Ezra Pound whose ideas influenced Eliot considered poetry "as a sort of inspired mathematics which gives us equations for the human emotions". Symbols help a poet to be as impersonal as a scientist. He rejected the 'sentimental and mannerish poetry' of the Victorian poets with their set-phrases, saddled adjectives piled upon each other the idle ornaments, mechanical rhythms and vague expressions without clinical precision in the descriptions. Both Wordsworth with his 'dull' descriptions of Nature and Tennyson with his long-winded statements are not safe models for budding poets. Pound's definition of a poem is startling as well as enlightening: "A poem is a product of extremely careful verbal engineering". Modernist poetry with its symbols are profound in conception, complex in structure and tightly woven. It contains no "verse essays" as in Wordsworth's *Prelude* or in Whitman's poems. Thus modern poetry is a "making" or verbal strategy and not "a spontaneous overflow of feelings". Both Pound and Eliot revolutionized poetic concepts and practice in the 1920s. Modern poetry is an exploration of the possibilities of language. Poetry's role became exalted by the use of symbols. They maintained a clear-cut distinction between functional prose discourse and a symbolic poem. They believed that the content was in the form and so no more should content be considered essential and form an accessory. In a good poet, change in technique always follows change in subject-matter. One of the axioms of modernism was "organic form"—form is never more than the revelation of content.

As modern poetry is an exploration of the possibilities of language, it reflects the social and moral problems of our time. A modernist poem aims at expressing the mind and emotions of the present day. Eliot and Pound are difficult and confusing because their poems have greater complexities of thought and intricacies of feeling than the reader has ever noticed. They aimed at effective expression avoiding the "fatal discursiveness" of the Romantic and Victorian poets. The poetry of Wordsworth and Tennyson suffer from long-windedness and misty vagueness. Genuine poetry must compress a situation into a single memorable image or symbol. Poetic concentration according to Paul Valery keeps language pure and dignified and in the Aristotelian conception of poetry it is "language made beautiful". Poetry keeps itself pure by "a refusal to describe or explain". In the words of Mallarme, "Poetry must never state anything directly but always give the pleasure of guessing". Poetry is language purified of the dross of day-to-day language. A symbolist poem cannot have a paraphrasable or translatable meaning. Form is no more conceived in a narrow sense to describe special 'metrical or stanzaic patterns' but in a wider sense of the sensuous imagery and auditory rhetoric of a poem. Thus the final value of a poem always springs from the inter-relation of form and content.

Post-war British Poetry

In the United Kingdom, the euphoria of winning a war was soon over as economic austerities were imposed. Britain became a post-imperial power. In the class-ridden British society; the rise of the working class and their expectations was a force to reckon with. In the 1950s a group of poets called The Movement expressed themselves in a way strikingly different from each other. The major poets were Ted Hughes, Philip Larkin and Thom Gunn. Gunn denied that label and later settled in America. The other poets were Kingsley Amis,

Robert Conquest, D.J. Enright and Donald Davie. The salient features of the movement poets are:

1. Anti-romantic
2. Witty
3. Rational
4. Sardonic

Unlike the Romantic poets The Movement poets were concerned about the propriety of making poetry. They held the belief that poetry should be well-made and craftsman like utilizing traditional rhythms and rhyme schemes. Poetry is not a matter of playing language games but making a verbal artifact. The maker cannot ignore quality. Poetry is *'logopoiea'* or word-making and a poet is a wordmaker. As a wordsmith working with the power of words he has to consider poetry as an affair of sanity and see things as they are. Most of these poets were admired for their clear thinking, simple expression and avoidance of strong feelings. They revealed openness to experience by providing psychological insights and were willing to face the realities of their time. Life is full of moving moments and they believed in the transforming power of poetry. The *'virtu'* of poetry which in Latin means power lies in its ability to capture the scenes of life and never let those special moments fade away. According to the poet and critic David Young, "A good poem is not a bundle of devices that add up to literature but an unerring trajectory into insight". The deconstructionist critic Paul de Man points out "the absolute randomness of language prior to any figuration of meaning in the act of reading poetry."

Ted Hughes

Ted Hughes was accepted as a classic of our time. His collection of poems has become set texts in schools and colleges. The critic Alvarez considers him as "a poet of the first importance". An association of poets in 1960's who came together to read and discuss each other's poems called themselves The Group. They considered Ted Hughes as a father figure and a dominant literary influence, for themselves.

They were Peter Porter, George Macbeth, Philip Hausbaum, Peter Redgrove and Edward Lucie Smith.

Ted Hughes and Philip Larkin differ most markedly in their choice of subject matter and in their handling of forms and have contrasting attitudes and powers. Hughes' best poems are about the animal kingdom. Some of his best known poems are "The Jaguar" "Thrushes" and "Pike". Though Hughes is fascinated by violence of all kinds that animals display he made his position clear when he remarked "My poems are not about violence but vitality". It is not glorification of violence for violence's sake but a celebration of power and the energy of life.

In the poem, *The Jaguar*, the hyper-activity of the jaguar within the confines of the cage is contrasted with the lethargy and boredom the other animals have accepted as a way of life. What follows is a formalist reading of the poem.

THE JAGUAR

The apes yawn and adore their fleas in the sun.
The parrots shriek as if they were on fire, or strut
Like cheap tarts to attract the stroller with the nut.
Fatigued with indolence, tiger and lion

Lie still as the sun. The boa-constrictor's coil
Is a fossil. Cage after cage seems empty, or
Stinks of sleepers from the breathing straw.
It might be painted on a nursery wall.

But who runs like the rest past these arrives
At a cage where the crowd stands, stares, mesmerised,
As a child at a dream, at a jaguar hurrying enraged
Through prison darkness after the drills of his eyes.

On a short fierce fuse. Not in boredom-
The eye satisfied to be blind in fire,
By the bag of blood in the brain deaf the ear-
He spins from the bars, but there's no cage to him

More than to the visionary his cell:
His stride is wildernesses of freedom:
The world rolls under the long thrust of his heel
Over the cage floor the horizons come.

In a formalist reading, close attention is paid to the form of a work. Formalists consider verbal investigation or analysis as the key to unlock the meaning of a work.

The verbs "yawn and adore" convey the sense of laziness as well as the unnatural behaviour on the part of the apes suggesting how captivity is killing them slowly. The tiger and lion are paradoxically "fatigued with indolence". Removed from their natural habitats and provided with easy meat the lords of the jungle are dying of boredom and lack of exercise. Death has claimed many of the animals in the zoo and those that are alive while away the time only by sleeping.

As a refreshing contrast to these lifeless forms resembling a painting on a nursery school wall, Hughes describes the jaguar as a supercharged piece of biological machinery. The jaguar refuses to "lie still" in the cage like the other wild cats. He refuses to accept his constraining circumstances. Hughes employs the "*bisociative*" technique to highlight the amazing vitality of the jaguar that spins like a top in the cage. The juxtaposition of the animate with an inanimate object like eyes and drills creates fresh perceptions. The jaguar is fired by the passion of a vision for freedom. Neither boredom nor the bars of the cage can hold him for long. His eyes are like drills that could cut open the iron bars of the cage. He is compared to a visionary who is already mentally free though physically a captive. The last two lines are highly characteristic of Ted Hughes. The hyperbole is used to create an intended effect. It is a celebration of the jaguar's power and untamed will. The poem is as much about the animal as about freedom.

The whole poem throbs with an explosive force as one image detonates another. Hughes believed that convention and hypocrisy of modern life have hidden the vitality which we ought to feel and which the animals and birds he admires still display. His poems offer valuable insights into the complete otherness of animal life. For human beings life is diminished or endangered by psychological distress. Hughes aims at educating human beings to be like animals in their best qualities. He tries to draw a moral for mankind from animals and birds. Hughes wrote: "The animals in my poems are in a state of energy which men only have when they have gone mad".

Philip Larkin

Philip Larkin is considered by many critics as 'the most important poet' after the war. He is more interested in the humanity of the people he sees around than in animals. He believed in calling a spade a spade. For him, art was the camera. Larkin's subjects include incidents, moods and moments of everyday life particularly about the hopes, fears and disappointments of modern man. His content and tone strikingly differ from Hughes.

The Whitsun Weddings is one of his best known poems. It is a masterly depiction of England seen from a train. In the poem, Larkin puts his thoughts across with astonishing accuracy.

> That Whitsun, I was late getting away:
> Not till about
> One-twenty on the sunlit Saturday
> Did my three-quarters-empty train pull out?
> All windows down, all cushions hot, all sense of being in
> a hurry gone.

Larkin writes with an amazing colloquial ease and has little time for poetic ornamentation. Reading Larkin's poems makes one recall Coleridge's famous distinction between the language of prose and poetry. Coleridge defined prose as "Words in their best order" and poetry as "The best words in their best order". This poem like several of his poems takes place on a journey:

> "All afternoon through the tall heat that slept
> For miles inland,
> A slow and stopping curve southwards we kept,
> Wide farms went by, short-shadowed cattle and
> Canals with floating of industrial froth;
> . . . hedges dipped
> and rose: and now and then a smell of grass

Larkin is losing himself in the excitement of sense-awareness during the journey. According to P.R. King, "This poem may be read as dealing with English society as a whole through its panorama of scenes and the couples who suggest the recurring life of a whole community, together with the traditional journey image as a symbol for life itself". The poem begins with Larkin looking at weddings with a negative attitude.

At first, I didn't notice what a noise
The wedding made...

Gradually, he looks at marriage with a positive view and sees it as a sort of social ritual that expresses a traditional sense of community and harmony.

Struck, I leant
More promptly out next time, more curiously.
And saw it all again in different terms:
The fathers with broad belts under their suits.
And seamy foreheads: mothers loud and fat;
...All down the line
Fresh couples climbed abroad: the rest stood around;
The last confetti and advice were thrown,
And as we moved, each face seemed to define
Just what it saw departing....

Larkin excels at describing human emotions as objects. His poems are devoted to analyze the wide range of human feelings in different situations. For him, poetry is a way of being honest. About his art of writing, he remarked, "I write when I feel strongly. I have no enthusiasm for obscurity".

His poem *Toads* is a complaint against 'the toad work' that squats on a man's life unless he is willing to relinquish his upper middle-class comforts and security. It is a poem that contains a clear-eyed, view of contemporary living and its problems with no illusions. Larkin's ironic tone has the detachedness of the artist. Man's dreams, hopes and ideals are relentlessly diminished by the reality of life. Poems of pure happiness are very rare in Larkin. Larkin's skepticism was the result of his being deeply influenced by Thomas Hardy.

Thom Gunn

Most critics unanimously declare that the best poetry written in England during the 1950s to 60s was written by the 'Larkin-Hughes-Gunn' triumvirate. Gunn was the youngest of the movement poets and much of his work is an attempt to come to terms with the nihilism of modern life. He has been hailed as a "modern metaphysical poet". His best poems have a compact philosophical elegance. Many of his poems express fairly complex ideas within intricately extended metaphors. His poems were often concerned with matters of love and passion, one of the traditional themes of the seventeenth-century metaphysical poets.

His poems explore aspects of the rift between thinking man and acting man, between body and mind, self and others, self and the natural world. He has a highly individualistic voice and is an explorer of identity. The divided self has formed the staple image of so much of Gunn's poetry. Gunn defined poetry as "memorable speech". He used rhyme and metre in a traditional manner.

Let us now attempt a Reader-Response reading of Thom Gunn's most famous poem *On The Move.* Reader-Response theorists consider the reader's interaction with the text as central to interpretation. No text according to them exists independently of the reader. The reader plays an active part in semantically weaving the text by filling in the gaps and drawing on his own life-experiences and knowledge.

A professional philosopher may interpret Gunn's *On the Move* from a Sartrean existentialist point of view and a college student in his early twenties may read the poem as mirroring his own spirit of adventure and search for meaning. Which interpretation is valid? Reader-Response theorists would consider both readings as possible and correct. The reader "creates" the work of literature during the course of reading. Meaning creation plays a more vital part than extraction of meaning.

The poem *On the Move* describes the aimless movement of a motorcycle gang without any sense of direction.

On Motorcycles, up the road, they come:
Small black as flies hanging in heat the Boys,
Until the distance throws them forth, their hum
Bulges to thunder held by calf and thigh.
In goggles, donned impersonality,
In gleaming jackets trophied with the dust.
They strap in doubt—by hiding it, robust
and almost hear a meaning in their noise.

The boys are not described as individuals but as machinery personified. The poem employs images from contemporary culture. The behaviour of the rebellious youths symbolizes our own generation in confrontation with our modern world. The youths feel boredom at the core of their existence.

They ride, direction where the tyres press
They scare a flight of birds across the field
Much that is natural, to the will must yield
Men manufacture both machine and soul
And use what they imperfectly control
To dare a future from the taken routes.

The threatening movement of the motorcycle gang becomes a metaphor for modern man's sense of alienation and lack of purpose. The boys camoflouge their real faces with goggles. They are searching for new values in a valueless world. The only satisfaction the youths have is that they are not idle. The images of "motorcycles", "goggles" and "dust laden clothes" bring out the robustness, careless spirit and urgency of the boys. On the surface, these images indicate action without purpose conveying the meaninglessness of much of human living.

At worst one is in motion; and at best,
Reaching no absolute, in which to rest,
One is always nearer by not keeping still.

The line, "men manufacture both machine and soul", in an existentialist interpretation will serve as the crux or punch line of the poem. The Existentialist philosopher Sartre taught how making one's soul involves the need for making choices at every stage of our life and taking responsibility for those choices. The poem ends on a positive note that in a world where traditional values are no longer the definite option, the only answer seems to commit oneself to action. By choosing to be on the move, the boys are at least getting away from something they find valueless.

This poem like most of Gunn's finest poems is marked by intellectual subtlety and complexity. He stresses on the power of the human will rather than reason as an active factor in man's role in a hostile world. It is considered his masterpiece and is marked by seriousness of thought and plainness of language.

Seamus Heaney

Seamus Heaney is one of the greatest Irish contemporary poets. He was influenced by Ted Hughes and Gerald Manley Hopkins. He won the Nobel Prize for literature in 1995. Much of his early poetry derives from his experience of the farm and country life as a child. He grew up in a rural landscape. Heaney's poetry is concerned primarily with the native, political, historical and cultural themes of Ireland. In some of his best poems one finds an exploration of his own relationship with the past history of his country. The political situation of his country has deeply affected his poetry. His work is notable for evocation of events in Irish history and its allusions to Irish myth.

The Death of a Naturalist is a collection of poems that deals with the loss of childhood innocence and the transition into adulthood. The two best-known poems from this collection are *Digging* and *Mid-term Break*. They deal with the poet's own inner struggle as a young man.

Digging is a poem with many themes. The poet articulates an inner conflict as to his choice of an occupation. He is certain that he cannot take his place among his father and forefathers as a farm labourer. He can honour the dignity of manual labour and the physical toil of his ancestors by choosing to write and preserve them in verse.

Between my finger and thumb
The squat pen rests
I'll dig with it

He had only two choices: follow his agricultural roots and inherit the farm or take up the pen as a tool mightier than the sword and work for reconciliation between the Republicans and the Unionists in N. Ireland. He had earlier rejected the voice of Republican militarism by refusing to take up the gun. The language of Digging with its verbal effects and evocative images are the unique trademarks of Heaney as a poet.

The cold smell of potato mould,
The squelch and slap of soggy peat
The curt cuts of edge
Through living roots awaken in my head.

In the above lines, the alliteration in the squelch and slap/ of soggy peat and 'curt cuts', the assonance of 'cold', 'potato', 'mould' and the onomatopoeia in 'squelch' and 'slap' echo the sounds they describe. Sensory descriptions are one of the strong characteristics of Heaney's poetry.

Punishment is a very disturbing poem and describes a photograph of a 'bog body'. It describes the fate of a young woman, who had most likely been shorn, stripped then killed and thrown into the bog as some form of punishment for a transgression unknown. In the opening stanza the poet expresses a sense of identification and empathy with the victim but soon becomes a voyeur and takes pleasure in the woman's exposed and naked body:

I can feel the tug
Of the halter at the nape
Of her neck, the wind
On her Naked front
It blows her nipples
To amber beads,
It shakes the frail rigging
Of her ribs.

The conflict of emotion indicated here is given further expression in the poet's direct address to the dead woman;

My poor scapegoat
I almost love you
but would have cast, I know
The stones of silence
I am the artful Voyeur

Here Heaney indicates that even though he is attracted to the woman he would have been complicit in her death by not raising support of her in protest at her punishment. Focusing the killings of the past with the contemporary conflict in N. Ireland and characterizes himself as

I who have stood dumb when your betraying sisters,
Cauled in tar wept by the railings.

In *Mid-term Break* he writes about the death of his younger brother and examines the theme of death in a literal and figural sense. The simplicity and linguistic economy of the poem makes his poems "say less and mean more" and carry the full weight of meaning. Critics have suggested that his poems represent a major achievement in twentieth-century poetry.

His appeal is immediate with the poetry-reading public because of his accurate descriptive powers and the clarity of his figurative language. The theme and language is always clear and straightforward. For example, the poem *The Mid-term Break* is built upon a simple extended metaphor. Death is compared to a break in the academic calendar of a college. The reactions of the people of various age groups to the death of his younger brother are described with sharp observation and unsentimental clinical precision. The last line "A four foot box, a foot for every year" is a marvellous example of linguistic economy—the implied sense of Death dogging the footsteps of a four year old child is filled with awe and mystery in a matter of fact manner. Heaney breathes a universal significance into his best descriptions. There is no elaborate philosophizing about the phenomenon of death. He does not write of ideas. The reality of death with all its individual particulars is conveyed in a direct language with no obscurity of expression. This style of writing gives his work powerful precision and lucidity. He always breathes a universal significance into his best descriptions.

In conclusion, it is relevant to highlight the contrasting views of the German philosopher Hegel and the English critic Philip Sidney on Art and poetry. For Hegel, philosophy is higher than Art because poetry is still tied to "picture thinking". For example, in a poem, snow may represent the idea of purity. Art is tied to "representation" (*Vorstellung*) or image-ination whereas philosophy deals with "Notion" or the realms of pure thought which is wholly conceptual. Truth proper is according to Hegel "imageless". The inner depths of the human mind elude full expression in art. Art is destined to be superseded by philosophy which is able to dispense with the sensuous and to deal in purely conceptual terms.

Philip Sidney considered poetry as immensely superior to philosophy, history and the physical sciences. Philosophy is highly abstract and general whereas history is concrete and particular. Poetry is superior to philosophy and history as it combines the merits of both. It is superior to the physical sciences; as the study of stars or matter has no effect on man's inmost being. Poetry works unlike the other disciplines upon the very centre of the soul, upon the moral consciousness of Man.

4

Unlocking the World of Drama

"Drama is but a re–enactment of life".—J.M. Synge

Drama is a performance art. It is literature written in a style that facilitates stage performance. In its text form drama is usually read as a story unfolding itself through the dialogues of the various characters and stage directions. Of all literary genres, drama comes closest to the language of everyday speech. Dramatists' convey their ideas through plot and characters while novelists and poets do so through narrators and personas.

A reader of drama must keep in mind that drama is essentially about life in motion and focus on the "dramatic" elements like conflicts and resolution, compelling actions and reactions. The reader must focus on the structure of the unfolding story and also on the inter-relationships of characters.

The structure of the plot is usually linear, unfolding from scene to scene, act to act. Some plays have a contrasting subplot to reinforce the theme of the main plot. The language of the characters in a play offers vital clues to their backgrounds, thoughts, feelings and unique personalities.

Understanding characters in a drama is more difficult than in fiction. The playwright does not inform the reader much about the characters directly. The reader has to carefully read the stage directions and make inferences about the characters from the dialogues.

It was Aristotle who defined the plot of a drama as "the structure of the incidents". It is described as a four-part movement:

1. Introduction
2. Complication
3. Conflict
4. Resolution

The introduction is the beginning of the action, the opening scene in which the major characters are introduced. The opening scene of a play is of great importance since it defines the situation from which the whole action will grow. The complication is the beginning of the conflict between characters for various reasons. The conflict reveals characters in opposition to each other or a reversal in the fortunes of a major character as in a tragedy. Finally, there is the denouement or the Resolution. It is the way the conflict gets resolved in the end. Comedies have happy endings and in tragedies, the protagonist is either exiled or dies leading to a sad ending.

An important thing to keep in mind when reading a play is that it was not meant to be read; it was meant to be seen and heard in one sitting of about two or three hours. Dramatic language has force and immediacy. Its primary function is also to create the 'setting', the world within which the action takes place. The value of great drama is based not on its being rooted in the particular and the historical but in universal human experience. This indeed is the value of all great literature too.

Dialogue in drama is an exchange between at least two characters. It gives the audience or reader the sense of the thought being formed by the character as he or she speaks under the pressure of a particular situation.

S.W. Dawson observes that "The nature of a play's world determines what can happen within the play, limits the possible situations, the range or depth of the action. These limits are what critics of drama used to call 'decorum'. There are many different kinds of limitation. Confining the characters and their speech to a particular social class is one very obvious way of defining a world—Ibsen's plays belong to a middle-class world and its own convictions". Thus the categorization of plays as comedies, tragedies, farces and so on is only to highlight the world the dramatist has created.

Every dramatist after establishing the limits of a play's world captures our attention by placing characters in specific situations. He further arouses our expectations by developing a tension in the actions and interactions of various characters. This is made interesting by the perfect naturalness in which the events unfold in the play. As the tension cannot be maintained for long, a resolution takes place with powerful effect. In some plays, suspense or mystery is presented in a very subtle manner. This is the supreme art of the dramatist in which the audiences are entirely uncertain of the outcome till the end.

Character is the life of drama. Through their characters great dramatists from Sophocles to Shakespeare reveal an insightful truth about the universal constants in human nature that cuts across socio-cultural and historical specificity. Both the classical Greek dramatists and Shakespeare teach us that certain human emotions and interactions do not fundamentally change. Drama has always been considered a serious literary form by the greatest literary masters of all ages and in all countries. The twentieth-century sparked a revival of interest in serious Drama pioneered by George Bernard Shaw called the New Drama. Shaw considered drama as "a vehicle for debating ideas about life". He was a socially committed dramatist. From Shaw to Stoppard, English drama has considerable significance for every reader. Most English critics from Dryden to Eliot have contributed to dramatic criticism on Shakespeare. Coleridge lavishly praised the dramatic qualities he found in Shakespeare. He considered Shakespeare as a writer of great plays than as a poet. He wrote: "In Shakespeare's poems the creative power and the intellectual energy wrestle as in a warlike embrace. At length in the drama they were reconciled". For greater analysis of Shakespeare's characters and his dramatic language we must turn to contemporary readings of Shakespeare.

Contemporary Readings of Shakespeare

Dr. Samuel Johnson one of the colossal literary figures of the eighteenth-century remarked, "Shakespeare lets you know almost everything about human nature". Why do we read Shakespeare in the first place? We never seem to wonder about the relevance of reading a sixteenth–century English dramatist to our times. Eliot asserted with utmost certainty that "Shakespeare will never be superannuated". No programme in Literature is complete without the study and appreciation of Shakespeare's plays. Shakespeare is an author whose plays are necessary for the growth of our minds. He deals with fundamental human problems that have plagued humankind from the dawn of its existence on earth.

The four great tragedies of Shakespeare are a classic example of his analysis of the human predicament from different angles. He never skirts the relevant issues nor deals with "pseudo-problems". In *Macbeth* he examines the painful consequences of ambition with no regard for ethics. Othello is a play that dissects the anatomy of suspicion—the most dangerous emotion to afflict marital love. In *King Lear*, the '*hamartia*' or tragic flaw is misjudgment. Lear's inability to assess the character of his children accurately precipitates his fall. *Hamlet* is a play that explores the serious consequences of vacillation or indecisiveness when it comes to decision-making on crucial matters.

Hegel described characters in plays as the motor of ideas. Every idea to be put into action needs a free human personality if it is to appear live and breathing before us. Shakespeare's genius precisely lies in making major human issues and problems inhabit characters and function through them. It is the individuality of his characters that contribute to the tragic sense in all the four great tragedies. As a critic jocularly remarked, "If Hamlet had been a Othello he would have killed everyone in the first act itself". If Othello had been a Hamlet, Desdemona would have escaped the brutal murder and the punch line of the play would have been "To kill or not to kill". It is the uniqueness of every character, portrayed as an "unrepeatable entity" which accounts for the unparalleled greatness of Shakespeare as a dramatist.

When most critics acclaim Shakespeare as the greatest dramatist in English literature, they presuppose certain standards of judgment. No writer can be exempted from these standards, which are indispensable for making value judgments. After reading the plays of Shakespeare, Ben Johnson and the 'mighty lines' of Marlowe, most critics consider Shakespeare as the best dramatist based on the range and depth of his works.

Today, Shakespeare is the world's compulsory author. He has received more commentary than any other writer. He is taught extensively in schools and colleges all over the English-speaking world. Shakespeare is essential for the liberal education he provides freely to every human being.

Postcolonial Reading of Shakespeare

Shakespeare's play, *The Tempest* can be read from the anti-colonial perspective. It is the favourite text of postcolonial critics who examine it to reveal attitudes to colonialism in Shakespeare's times. Prospero is seen as a symbol of the white colonial ruler who enslaves the "savage" Caliban and Sycorax, the indigenous people of the island. *The Tempest* serves as a happy hunting ground for not only a re-thinking of Jacobean imperialism but of all British colonial operations in Third World countries. Thus *The Tempest* reflects "the imperialist rape of the third world" for a postcolonial critic.

In postcolonial readings, Prospero becomes 'the universal father figure of pedagogy' and 'the prototypical capitalist usurper'. Ania Loomba brilliantly demonstrates how the play reveals the consequences of colonization on the natives. Colonization is the manifestation of unequal power relations and the relationship between Prospero and Caliban is a master-slave struggle. Caliban is the betrayed innocent savage or the "speaking sub-altern". In the power struggle Caliban is defeated and weakened and easily becomes the manipulated tool of the oppressor.

Prospero wants Caliban to 'mimic' him after learning his language but Caliban as a typical postcolonial subject shows his resistance by mocking him. Caliban is portrayed by Shakespeare in terms of the savage *"Other"*. Edward Said comments on how the language of Caliban was changed into the Master's language—a typical instance of neo-colonialism. Shakespeare depicts him as possessing no language of his own and therefore deserved enculturation by means of colonial domination.

Liberal Humanistic Reading

Brain D. Vickers in *Appropriating Shakespeare* maintains that Prospero in *The Tempest* is a representation of "Divine Providence on the island". He ushers in prosperity on the island. Caliban the unfettered savage inhabits the natural elements of Earth and the Woods. Ariel inhabits water and air. Prospero a God-like figure controls all the accidental elements of nature to bring his enemies under his control.

Homi Bhaba mocks at this interpretation as a typical colonial interpretation of the play. The omniscient (All knowing) White Prospero 'discovers' the new lands controls 'Air' (air force), 'Water' (navy), the 'Earth' (guns and artillery) and destroys the 'Woods' (natural resources). It is the greed for territory and material possessions that fuels the colonial expansion.

New Historicist Reading

There is a definite political turn in Shakespearean studies today. Critics have moved from character study and textual analysis to new historicist and cultural materialist readings. The term "New Historicism" was coined by the American critic Stephen Greenblatt in 1982. A New Historicist reading of a Shakespearean play like *Macbeth* will focus on the text as part of a network of cultural forces, linking 'literary' texts with 'historical' material like the role of political or religious authority in Jacobean society. The entire social and cultural history of that period is considered to be embedded in the literary text. A sub-text in the play like the persecution of witches would be analysed with all the related historical documents. Both, the play Macbeth and the historical texts are given 'equal weighting' and the boundary between the literary text and the historical background is erased and everything becomes text. The play cannot be read in isolation because of the "*social embeddedment*" and has to be along with the historical text.

The point is to reveal the awe in which a monarch was held in Renaissance England. The play is assumed to explore and justify "regal ideologies" and to show how Shakespeare was part and parcel of his society. The most convincing proof is the "appropriation" of Shakespeare's troupe by King James I who declared them the "King's Men". Thus a New Historicist reading explains how literature and religion are various forms of cultural expression that make up a society and endorse its values at specific historical times. Literary Criticism thus becomes culture criticism.

Cultural Materialist Reading

Cultural Materialism is heavily influenced by Marxist ideology and is different from New Historicism in its focus on the present than on the past. In a cultural materialist reading of Shakespeare, textual analysis is combined with historical context, theory and political commitment. Literary criticism is considered as a form of political resistance both in condemning the past and 'challenging' the present. Everything is reduced to politics or ignored. It situates Shakespeare's plays within our own political situation.

A cultural materialist reading of Shakespeare's *Merchant of Venice* will focus on questions of class ideology and economy with a firm and explicit political commitment to the transformation of a social order that exploits marginalized groups on the grounds of race, gender and class. Thus literature is either an emancipation from social oppression or it is collusive with power.

Alan Sinfield's essay *Cultural Materialism: Othello and The Politics of Plausibility* is a brilliant example of a cultural materialist reading of Shakespeare's *Othello*. Othello is one of the most widely discussed plays of Shakespeare. Othello by race is a Black Moor. The "symbolic geography" maintains the conscious division between the White race and the "*Other*". The Other signifies evil, brutishness and monstrosity.

When Desdemona speaks about her 'divided duty', she is posing a challenge to both Patriarchy and Racism. By marrying a Moor, an outsider, against the wishes of her father, Desdemona poses a threat to patriarchy. By marrying outside the Venetian nobility she defies oligarchy and by marrying a Black man she strikes a death blow to racism.

The pertinent question is whether she is totally free from racist attitudes. Her inner struggles with racism at the sub-conscious level become evident when she says "I saw his visage in his mind". She claims that though Othello is Black, he is still attractive. From the conservative point of view, Desdemona as a white woman is a symbol of absolute purity and beauty.

Emilia on the other hand is an example of an abused woman. She suggests the need for a home for battered wives. Bianca belongs to the class of "impure women". Bianca in the words of Iago is a woman who by "selling her desires, buys herself bread and cloth". She is a prostitute and represents the marginalized in the play. Thus, women are portrayed as soft targets and victims of suspicion and oppression in the play.

As the distinguished Indian novelist Shashi Deshpande observed, "Of what use is art if it is not wedded to a social purpose"? This social transformation can be achieved only with good art. Cultural Materialism's main preoccupation is with the present rather than the past. The contemporary Marxist critic, Terry Eagleton asserted that, "Shakespeare is the quintessential commodity at once ever new and consolingly recognizable, always different and eternally the same".

Feminist Reading of Shakespeare

A Feminist approach to Shakespeare would examine a play like *Romeo and Juliet* or *The Taming of the Shrew* to reveal how patriarchal power and gender relations are worked out in Shakespeare. *King Lear* would be considered as a representation of patriarchal misogyny and offering lessons in family responsibility. The conclusion drawn would be bleak.

None except those who endorse misogyny can endorse the ending.

Though Shakespearean criticism of the earlier centuries held an established view that Shakespeare has only heroines in his comedies and not heroes, twentieth-century Feminist criticism emphasizes the victimization and exploitation of women in the plays. In *Othello* for instance, men are the villains, controlling, cheating and destroying women. The status and perception of women in Renaissance England are highlighted in a feminist reading.

The Problem of Shakespeare's Language

Shakespeare can be hard to read as the English language has changed in the last 400 years. The vocabulary and syntax used in Shakespeare's times was completely different from modern English. Some words like 'beshrew' which meant 'a curse upon' and 'cantelous' which meant 'crafty' have vanished from the English spoken today. Certain words like 'clip' and 'nice', which meant 'embrace', and 'shy' in Shakespeare's works have different meanings in contemporary English. Modern readers of Shakespeare are often confused by such changes in word meaning but by relying on a good modern edition with clear explanatory notes on the same page there will not be any difficulty in understanding Shakespeare's text. Some scholarly editions that will enable us to enjoy Shakespeare are the Oxford Shakespeare, the New Cambridge Shakespeare and the New Arden Shakespeare (1995 edition onwards).

In conclusion, one can agree with Jan Kott who in his book *Shakespeare: Our Contemporary* reads Shakespeare in order to demonstrate not only Shakespeare's universal genius but also to show his continuing relevance to our times. Kott argues how Shakespeare could offer immediate solutions to current political affairs as he is immortal and as relevant today as he was in the sixteenth century.

Samuel Beckett

Samuel Beckett's *Waiting for Godot* one of the cornerstones of modern drama, is a prolonged and sustained metaphor on the nature of modern societies. The idea of Godot as a play in which "nothing happens, twice" is to express the boredom, ennui and "the suffering of being". Beckett says at the end of the play that the problem was how to pass the time and see life through. Beckett dispenses with plot and background, as his characters are those without roots. Beckett declared about his own play in the following words:

> We have no elucidations to offer of mysteries that are all of their (critics) own making. My work is a matter of fundamental sounds (no joke intended) made as fully as possible, and accept responsibility for nothing else. If people want to have headaches among overtones let them and provide their own aspirin.

The Impact of Beckett's Plays

Backett is an iconoclast and image-breaker. He has shattered conventions and pioneered a new kind of drama. It is rightly said, "His drama does not teach lessons, it does not preach, it does not do any propaganda. His plays show the situations in which we all are; they expose us in our existential predicament".

Beckett's plays leave their indelible expression on the minds of their audiences, because they challenge some of their assumptions about life and living. One of the critics observes: "Beckett's plays stay in the bones. They haunt me sleeping and waking coming upon me when I am least aware".

Beckett communicates an experience. The experience of words and non-words. He once remarked that "Speech is a stain on silence". He expresses his vision of desolation with unique power.

The Characters in Beckett's Plays

Beckett's characters are without a past or future. They are timeless creations. They exist as long as the play lasts. Beckett gives us no hint as to how they have come to the situation in which we find at the end of the play, they remain unchanged. Beckett has an awareness of man's solitude, imprisonment, and pain.

Existentialism in the Play Waiting for Godot

Jean-Paul Sartre defined existentialism as the doctrine that "Existence precedes essence". He distinguishes between inanimate objects, or a "being-in-itself" and human consciousness a "being-for-itself".

The character of inanimate object has been assigned to it. On the other hand Sartre claimed that if one is happy it is by their own free choices. In this sense, humans exist and then define and choose their essence. Existentialism is a revolt against traditions and conventions and against all that hampers human liberty. It is a movement for individuality, for human freedom and human rights. It is an advocacy for man's existence as man with no external barriers and laws. As applied to literature, 'Existentialism' emphasizes man's responsibility for forming his own nature as well as his personal decisions, personal freedom and personal goals. Man is entirely responsible for himself.

A Brief History of Existentialism

Most people believe that Soren Kierkegaard (1813-1855), Gabriel Marcel, Jean Paul Sartre and Albert Camus (1915-1960) are the chief exponents of 'Existentialism'. The Danish philosopher Kierkegaard raised his voice against Western materialism, industrial revolution, scientific research and machine culture and the modern evils that were haunting and darkening human existence. He propagated Existentialism against Hegel's idealism, which said, "The real is the rational and the rational is the real".

And Sartre was interested in the nature of existence, freedom, consciousness and time. Highly influenced by the work of Soren Kierkegaard, Friedrich Nietzsche and Edmund Husserl, Sartre helped to develop a philosophical Movement called Existentialism.

The Major Beliefs of the Existentialists

According to them, man is because man exists. Man's existence is essentially miserable, lonely, alienated; their desire for happiness is a myth, and there is no escape of man from suffering. And this is what we find in the play; *Waiting for*

Godot. The Existentialists regard human life as absurd. They believe that the individual is the product of the choices he makes.

In literature, existentialists have highlighted man's alienation, loneliness, suffering, helplessness, freedom, absurdity of human existence, etc. To them, man is an isolated being who is cast ignominiously into an alienated universe moving from nothingness to nothingness in a world that is waste, anguished and absurd. They believe that the individual is the product of the choices he makes. What baffles us most in reading this play is the identity of Godot. He is a mysterious personality. We often ask ourselves "Who is Godot?"

Waiting for Godot is a difficult play; It's meaning cannot be explained easily. When asked what his play meant, Beckett replied: "If I could tell you in a sentence I would not have written the play".

When asked what Godot represented, Beckett answered quite simply,

"If I knew, I would have said so in the play".

The two tramps wait for him, but Godot seems to be a kind of distant mirage. At the same time, the two main characters in the play, namely, Vladimir and Estragon feel uneasy about Godot. When the time comes to meet him they will have to approach him "on their hands and knees".

Waiting for Godot can be interpreted in many different ways. There are lots of Christian allusions but each reader can read it according to his own beliefs or non-belief.

Although Godot fails to appear in the play, he is a real character as any of those whom we actually see. However, the subject of the play is not Godot; the subject is "Waiting". The act of waiting is an essential characteristic aspect of the human condition. Throughout their lives, human beings always wait for something and Godot simply represents the "Objective of their waiting—an event, a thing, a person, death". Beckett thus depicted in this play a situation, which has a general human application.

The Themes of Waiting, Ignorance, Impotence and Boredom

As the play moves forward, we are made to realize that Vladimir and Estragon are waiting and that their waiting is of a particular kind. "Although they may say that they are waiting for Godot, they cannot say who or what Godot is, nor can they be sure that they are waiting at the right place or on the right day, or what would happen when Godot comes" thus they cannot get the essential knowledge, and they are ignorant.

Without the essential knowledge they cannot act, and so they are impotent.

And Vladimir and Estragon understand it perfectly. For instance, "Come on Gogo", pleads Didi breaking off on the rejection of one of the two thieves crucified with Christ, "Return the ball can't you once in a way?"

As a critic says, "Beckett in his drama does not write about things but presents the things themselves".

The Mood of Vain Expectancy

Vladimir and Estragon, have travelled towards total nihilism, but they have not fully achieved it. They still retain enough remnants of hope to be tormented by despair. And in the place of hope as a dynamic, they move expectancy. This is the main motif of the play, spelt out in the title which in an earlier version was simply: "Waiting".

They have to remain where they are, though they resent doing so and would like to leave. If Godot comes, a new factor may be introduced into their existence, whereas, if they leave they will certainly miss him. This mood of expectancy has also a universal validity, because whenever we wait we are expectant even though we are almost certain that our waiting will not be rewarded. This produces "suicidal impulses" in the characters. The characters have different nationalities thus symbolizing the universal human condition. For example, Vladimir is Russian, Estragon is French. Lucky is English and Pozzo is Italian.

"Waiting" and "The Flow of Time"

It is in the act of waiting that we experience the flow of time in its purest, most evident form. Being subject to the flux of time, human beings are, at no single moment, identical with themselves. For example, when Pozzo and Lucky first appear, neither Vladimir nor Estragon seems to recognize that Estragon even takes Pozzo for Godot, because after they have changed since their last appearance. Estragon even insists that he did not know them while Vladimir insists: "We know them, I tell you. You forget everything".

To wait means to experience the action of time, which is constant change. And yet, as nothing real ever happens, that change is in itself an illusion. "The more things change, the more they are the same. That is the terrible stability of the world". "The tears of the world are a constant quantity," says Pozzo, "for each one who begins to weep somewhere else another stops".

Waiting for Godot is not about Godot or the nature of Godot but about the waiting for him. It is about life on earth not the hereafter. The essential theme is how to endure life. Life becomes unbearable without meaning and purpose. Human kind tries to make it bearable through hope as an antidote to despair. In the words of Husserl, "To trust is a natural human disposition". Thus, we continue believing that there is some point to existence, we have to keep some mysterious appointment on earth and that we are not random scraps of life, is what is symbolized by the patient waiting for Godot.

The Pointlessness of Existence

This play is a parable. Godot may stand for God, or for a mythical human being, or for the meaning of life, or for death, or for something else. The play is a fable about a kind of life that has no longer any point. This fable is a representation of stagnant life. It is a fable that suffers from a lack of cohesion because lack of cohesion is its very subject matter. This fable offers no story, because it describes man eliminated from, and deprived of history.

It is meaningless to ask who or what the expected Godot is. Godot is nothing but a name for the fact that the life which goes on pointlessly is wrongly interpreted to mean waiting, or as waiting for something. What appears to be a positive attitude of the two tramps amounts to a double negation: Their existence is pointless and they are incapable of recognizing the pointlessness of their existence. (Beckett himself said that he was not so much concerned with Godot as with waiting).

There is sufficient resemblance between the standpoint of Beckett and that of the Existentialists to warrant a comparison of the two to study *Waiting for Godot* as a play having Existential elements.

The protagonists are placed in a very unhappy predicament; "They have to live like serfs". The world is full of boredom, meaninglessness, and life is without purpose. Man may wait ceaselessly for the end of his suffering and misery, yet he would find that there is no end for human misery. The two tramps produce in us a sense of baffled helplessness, which we experience when forced to remain in a situation, which is beyond our control. The condition of the two tramps is hopeless, because Estragon suddenly offers the suggestion: "Let us hang ourselves immediately". In the words of John Barth, "When to live has no reason, there is no reason to live".

The tragic quality of the play is deepened by the Pozzo-Lucky interludes, Lucky's plight is indeed pitiable. He is no better than a beast of burden, and there are sores on his neck. Pozzo-Lucky episode shows that the entire human race lives in a universe, which is not looked after by an exactly kind the God. The characters are in anguish and full of pain and suffering have the existential tendencies, as also the theme is existential.

Waiting for Godot deals with the theme of despair. It deals with man who continues existing because he happens to exist. But however, the play is not a traditionally existential play in spite of their inaction and pointlessness of their existence, the two tramps still want to "Go on". Ultimately they go on living, because they happen to exist and because existence does not know of any other alternative but to exist.

Vladimir and Estragon may be regarded as brothers who are deprived of history. They are like men who never having been married continuously expect the return of their wives. And in Beckett's eyes we are all like them. From the beginning to the end, Beckett asks one question. "When there is nothing to express, nothing from which to express, no power to express, together with the obligation to express, can language be an adequate substitute for the great fog of boredom that envelops every event of our times". The critic Susan Sontag calls it "the rhetoric of silence".

Tom Stoppard's *Jumpers*

Tom Stoppard was one of the greatest post-war British playwrights. He was famous for writing plays with a philosophical subject matter. He embraced absurdism and his specific objective was to blend the play of ideas with farce. In *Rosencrantz and Guildenstern are Dead* (1966) he wrote a witty comedy by foregrounding minor Shakespearean characters. In *Jumpers* (1972) he defends liberal humanism and its values, which was under attack in the universities. Stoppard's plays never fail to delight us with their juxtapositions and puns. In his play *The Real Thing* (1982) Stoppard makes a distinction between a good play and a bad play. In the middle of the play, he compares a good play to a cricket bat hitting a ball. A good play can hit ideas and a bad play is a cudgel that can't hit farther than a lump of wood. It is a seminal play exploring the nature of drama. Henry, one of the major characters in the play is a playwright. Only in scene II, you realize that scene I is a performance from Henry's play and not part of the real play. It is a play within a play raising all sorts of questions. Stoppard has got wit; sharp and polished dialogues that are brilliant like what we find in the plays of Congreve and Wilde.

Jumpers is a philosophical play published in 1972. In this abstruse play, the pernicious influences of a major philosophical system of thought in the twentieth century is exposed. Logical Positivism, on human attitudes and conduct is analyzed in terms of a collision between theory and reality, the nature of good and evil and the problem of values.

Jumpers centres round three major characters, George Moore, an eccentric professor of philosophy, Dorothy, his nervous wife and a former actress and Archie, the Vice Chancellor of the University where George lectures. It is essentially a "play of ideas" in which Stoppard sets up George and Archie as the protagonists who provide the necessary intellectual pyrotechnics for and against the philosophy of Logical Positivism. Under the guise of farce, it is indeed a deeply moral play. In fact, Stoppard was defending the values of liberal humanism which was under attack by the logical Positivists.

Logical Positivism is a school of philosophy which originated in Vienna in the early 1920's and claims that a statement has meaning if it can be empirically proved to be true or false, and which aims to develop formal methods, similar to those of mathematics for carrying out this kind of proof. It recognized only the positive sciences as valid sources of human knowledge with its verification principle and rejected all traditional areas of metaphysical, ethical or aesthetic enquiry as meaningless as it was not directly verifiable.

In England, A.J. Ayer, a philosopher at Cambridge was the leading exponent of Logical Positivist Orthodoxy. For the Logical Positivist, ethics has no meaning since we cannot 'know' goodness in the sense in which we can 'know' that grass is green. Therefore statements which express moral judgements are pure expressions of feeling and therefore do not come under the category of truth and falsehood.

The major champion of Logical Positivist Orthodoxy in *Jumpers* is the University's Vice Chancellor, Sir Archibald Jumper, who is a man of varied talents in the fields of philosophy, law, psychotherapy, politics and gymnastics. The play's title is a pun on his own name and that of his students who are "jumpers", a troupe of gymnasts performing on a stage to celebrate the victory of their political party at the elections. As they were assembling themselves into a human pyramid a gunshot is heard resulting in the death of George's colleague and philosophy professor Duncan McFee.

After the murder, the mystery deepens when Dorothy tries to dispose of the corpse in a huge plastic bag. She is rescued from the clutches of the law by Archie, whose relationship with her is suspect. When Inspector Bones arrives at her flat for an investigation, Archie resourcefully gets rid of both the body and the Inspector. "Archie says" is a recurrent phrase of Dorothy in the play which indicates the tremendous amount of influence he has over her. Archie's statement, "No problem is insoluble given a big enough plastic bag" (p. 41), is not just absurdly comic given the manner in which Archie throughout the play highlights the moral limbo which is the consequence of Logical Positivism's rejection of all moral and spiritual values and value-statements as meaningless and illogical expressions of feeling. For Stoppard, Archie the sole dedicated representative of Logical Positivism is the villain of the play, for whom corruption is all right provided it works. Archie says:

> The truth to us philosophers, Mr. Crouch is always an interim judgment. We will never know for certain who did shoot McFee. (p. 81)

George is the play's hero and the spokesman for humanist values. His remarks about Archie to Inspector Bones in the course of the investigation of McFee's murder is sarcasm at its best.

Bones: Tell me something—Who are these acrobats? George: Logical Positivists, mainly with a linguistic analyst or two...a mixture of the more philosophical member of the university gymnastics team and the more gymnastic members of the philosophy school. The close association between gymnastics and philosophy is I believe unique to this university and owes itself to the Vice Chancellor, who is of course a first-rate gymnast, though an indifferent philosopher (p. 41).

This description of Archie as a philosophical gymnast, apart from the joke suggests the dangers of Logical Positivism upon the minds of those influenced by it. Human values are seen to be corroding as a direct result of such materialistic habits of thought. Any kind of amoral or unprincipled conduct can be explained away by reducing all philosophy to narrow logical premises and carefully worded assertions in which all utterances about morality are nonsensical. The murder of McFee and the haunting question of '*Whodunnit*' is the central event on which Stoppard builds the interplay of ideas for and against Logical Positivism.

To conclude, Stoppard shares with Bernard Shaw the idea that a play can be a vehicle for debating views on life. Stoppard's warning is about the decline of established human values with the aggressive onward march of materialism in our postmodern societies. He alerts us to the dangers of the adverse consequences of such a debasement of ethics through the character of George Moore who as a determined humanist attacks the intellectual postures of a Logical Positivist like Archie for whom truth is provisional and generosity and selflessness are merely emotional whims. Stoppard, through George as his mouthpiece tries to demonstrate the value of the non-verifiable content of life and to prove that there is a source of meaning and values in the universe.

5

The Linguistic Universe of Modernist Fiction

> "A book ought to be an axe to break the frozen sea within us".—Kafka

The novel continues to be the most popular literary form even in the Digital Age. Most of the noble laureates in literature are novelists. The novel as a literary genre has its origins in eighteenth century England. Ian Watt in his book *The Rise of the Novel* traces its origins to the growing emphasis on the importance of individualism as a social value after the industrial Revolution and the rise of the middle-classes. The interest in the lives of ordinary people in the eighteenth-century, stressing the importance of the common man as opposed to kings, nobles and generals in Shakespeare's times was a crucial reason for the novels rise.

Mark Spilka, an American critic observed that "The worth of individuals in newly democratic societies is very much the novel's subject and its raison d'être, and as such, crucial to its initial wide audience appeal and its continued appeal through the nineteenth-century and perhaps even today". Fiction is popular as it explores different facets of human nature and enables readers to identify with characters in their conflicts and visualize life with greater clarity. Reading fiction is a way of responding both emotionally and intellectually to the complexities of life. A great novel like Charles Dickens'

David Copperfield or D.H. Lawrence's *Rainbow* enlightens, humanizes and sensitizes us to one's actual living experience.

Raymond Williams, a foremost cultural critic of our times, is lavish in his praise of the novel as the most complete genre. In his essay, "Realism and the Contemporary Novel", he writes: "Now the novel is not so much a literary form as a whole literature in itself. Within its wide boundaries, there is room for almost every kind of contemporary writing.... When I think of the realist tradition in fiction, I think of the kind of novel which creates and judges the quality of a whole way of life in terms of the qualities of persons. It is what War and Peace does; what Middle March does; what the Rainbow does".

The novel as stated earlier developed in the early eighteenth-century England with the rise of an independent middle class. Williams identifies the major tradition of European fiction in the nineteenth-century as a tradition of 'realism'. He defines this realism as describing elements of ordinary everyday experiences, objects, scenes, external aspects of persons, etc. Engels defined 'realism' as "typical characters in typical situations". For Williams, Realism is not an object, to be identified, pinned down, and appropriated. It is, rather a way of describing certain methods and attitudes. The most common definition was in terms of an ordinary, contemporary, everyday reality. Realism is not idealism or caricature but describing a situation in a faithful, accurate, life-like manner.

Raymond Williams identifies many types of realism in contemporary fiction. The conventional realistic "social novel" he says has been replaced by the "psychological novel". In the works of novelists like Arnold Bennett and even the immensely popular Jane Austen everything was mostly presented from the outside and the inner reality of the persons is conveniently ignored. In the 'psychological novel' of Virginia Woolf and James Joyce, 'realism' becomes 'fidelity to psychological reality'. It is a different kind of realism that attempts to capture the inner reality or "felt reality" of our minds through "the stream of consciousness technique". Woolf's *To the Lighthouse* and Joyce's *Ulysses* are the best examples.

The modernist fiction of Woolf and Joyce was influenced by the ideas of the French philosopher Henri Bergson. Bergson's seminal ideas were explained in his books *Time and Freewill* and *Matter and Memory*. These ideas influenced the stream of consciousness technique in literature. Real time for Bergson is biological time, an inner duration that can be felt not analyzed. Real time is a defiance of all chronometric values. Clock time is a materialistic idea. It is a spatial segmentation of homogenous points placed side by side on a clock dial. Real time is growth, change and development. Time as the mind experiences it, is subjective not objective, kaleidoscopic and not linear.

Time as experienced by our human consciousness is not governed by the calendar or clock time. Time in the inner human consciousness is a duration, a flow. Bergson in his theory of time disputed the validity of chronological time and dismissed the concept of clock time as a human construct.

Measuring time in terms of seconds, minutes and hours and defining the passage of time in terms of the tick-tock of the clock is to lose sight of experiential time. The human consciousness experiences time as the present moment which contains the past as memory and the future as desire or vision. The present or the Eternal Now (St. Augustine's term) is no more than a meeting point in which the past and future coalesce in the Living Present (Husserl's term). Thus for Bergson, Time is not only eternal but internal. Man does not live in time, but time lives in man. Time in the hand which is clock time or External time is not to be in control of time, as the poet Adrienne Rich observed.

Thus Woolf and Joyce departed from the earlier realism of popular fiction writers like Arnold Bennett, John Galsworthy and Anthony Trollope. They ignored external reality and logical structure in their fiction. In their experimental fiction there was a collapse of the spatio-temporal continuum. There was no movement from point to point. The Aristotelian principle of causality in plot structure, the notion of beginning, middle and end was rejected along with descriptions of external objects and surface appearances. Superficial realism was replaced with 'felt reality' and a non-linear narrative punctuated by flashback and flash-forward in time. The earlier realistic novel "went out with the hansom cab".

Williams calls this type of psychological realism as "an equally damaging unbalance" as the socio-cultural space in which individuals operate as social beings simply recedes or totally vanishes and "we are left with voices and feelings, voices in the air". Woolf and Joyce can be called as pioneers of an experimental fiction but according to George Lukacs the Marxist critic, their ignoring the socio-cultural context makes their fiction solipsistic. By being out of touch with external reality, alienated from social life and because of the very low intake of human stuff they were incomprehensible to the readers.

What is needed is a balanced description of personal life (turning inwards) and depicting the quality of general life (the outer life) in a novel. This achievement of a balance between social and personal reality is evident in D.H. Lawrence's *Rainbow*, Tolstoy's *War and Peace* and George Eliot's *Middle March*. In all these novels there is "a valuing of a whole way of life" in Williams' assessment.

Reading D.H. Lawrence's Novels

D.H. Lawrence is considered as a dominant interpreter of the role of sexuality in our culture. His novels both illuminate and reflect the crisis of our society. Lawrence's distinction as one of the greatest realists in literature is the reflection of a socio-cultural context or mileu in his novels in which personal relationships are admirably interwoven with the larger society outside and never reduced to a mere scenic background.

Dr. F.R. Leavis includes D.H. Lawrence within the "Great Tradition" of English fiction along with Jane Austen, George Eliot, Henry James and Joseph Conrad. These five novelists in his assessment not only changed the possibilities of the art but promoted human awareness of the possibilities of life.

David Herbert Lawrence is considered as one of the most thought provoking novelists of the twentieth century. His greatness and relevance as a writer is seen not only in his critique of modern industrialism which leads to a kind of "death-in-life" existence but also in his presentation of an alternative vision of "real living" which was possible only through a balanced man-woman relationship in which the body and mind act in unison.

Lawrence insists with a moral intensity and a vehement urge the importance and necessity for human beings to become "Man alive" both in body and in mind and cease functioning in fragments. This was the only potent antidote he prescribed for the crippling effects of increased mechanization on the human psyche in our "tragic age". His major concern was the attainment of wholeness of being by recognizing the importance of sexuality in human existence and its right expression in the context of a committed heterosexual love-relationship. Lawrence had a reverential attitude to sex as a privileged and integral expression of love between a man and a woman. He never advocated promiscuity in his writings.

His philosophy of a "full spontaneous being", the possibility of which lay in the dynamics of an intimate man-woman relationship lies at the heart of his art. In contrast to "the Lesser Novels" like *Aaron's Rod, Kangaroo* and *The Plumed Serpent* which only reveal his ideological explorations of the leader-cum-follower idea and the political process; the principal novels like *Sons And Lovers, The Rainbow, Women In Love* and *Lady Chatterley's Lover* contain the core theme of all his writings which is the relation between man and woman. His life-long mission as a writer was to help men and women realize the importance of their corporeal self in the attainment of wholeness of being. It is precisely the reason for limiting the scope of this study to a detailed examination of Lawrence's four major novels.

He continues to provoke and disturb his readers even today with his visionary ideals of living and unique concepts of sex in human behavior. His novels also highlight a refreshingly new approach to one of the most intimate forms of communication between a man and woman. He sees sex as an integral aspect of our living and loving as human beings. His writings are permeated with a moral concern to show humanity the way to discover a whole new life. Though Lawrence was unfairly denounced as a pornographer in his times, his attitude to sex is one of respect and great awe at its unfathomable mysteries. It is akin to religious attitude to sacred matters. He was never a libertine in his personal life nor did he ever encourage promiscuous behaviour. He condemned pornography, considered marriage sacred and lifetime commitment to each other as the proof of real love. He never failed to express his shock at the depravities of people who displayed permissive attitudes and behaviour.

Lawrence's concept of "Real Living" with relation to the dynamics of gender relationships, is not an abstract philosophy or a metaphysical system but an exploration of the possibility of achieving complete fulfillment in our lives through the inner dynamics of a holistic man-woman relationship.

In *Sons and Lovers* the relationship between Walter Morel and Gertrude his wife portrayed with a tonic sense of realism. Their marriage foundered as a result of total incompatibility. Mrs. Morel was an educated, cultured and refined woman from the English middle-class whereas her husband was a coalminer who constantly exhibited the rude and crude characteristics of the working class in his conduct. Initially it was his irrepressible vigour and physical attractiveness, which draws Gertrude to him. Gradually when the romantic haze began to wear off she became deeply disillusioned with him as there was nothing endearing and substantial at the back of his show. Lawrence is careful to point out how physical attraction alone is a fragile foundation for holding together a marital relationship. It was a tragic failure of the purely sensual correspondence without the essential meeting of minds.

In the case of Miriam and Paul, it is Miriam's "passional paralysis" that finally leads to the breakdown of the relationship between them. She is a cerebral type of woman who cannot accept sexual love as a bodily reality. Paul terminates his relationship with her out of intense frustration as her denial of the sexual roots of love creates a barrier in the attainment of a satisfying and holistic relationship between them.

With Clara Dawes, it is the inverse that takes place. Paul becomes obsessed with the grossly physical and makes no attempt to connect to her as a person with emotional needs. This chronic deprivation of attention and affection on Paul's part forces Clara to terminate their affair abruptly. The isolated sexual acts without the commitment of a loving relationship make her feel used and she wanted to save herself from cheapness.

Lawrence implicitly diagnoses the causes for the three failed relationships. It is primarily due to the characters operating from two diametrically opposite forms of consciousness without making any sensitive adjustments to understand and accommodate each other. It is the case of the purely carnal self-indulgence pitted against the cerebral or spiritual consciousness and vice versa.

In the *Rainbow,* Lawrence describes the love and marital relationships of the Brangwen family spanning three generations. The relationship between Tom Brangwen and his Polish wife Lydia Lensky illustrates how only realistic love can succeed in the long term whereas romantic love with its inherent self-deception can leave couples completely disillusioned with each other very soon. After the initial difficulties in their marriage because of cultural and temperamental differences their marriage succeeds as the couple learns to accommodate each other's interests and learn to tolerate minor character flaws. The relationship between Tom and Lydia is presented by Lawrence as the most satisfying and successful of all three man-woman relationships in the novel.

In sharp contrast to the tenderness and sympathetic understanding of each other which is the hallmark of Tom and Lydia's relationship the relational dynamics between the second generation Brangwens, Will and Anna is marred by a tempestuous battle of wills. Their marital discord and the constant friction between them is the result of a deep-seated incompatibility between them as a couple.

Will is deeply religious and a creative person whereas Anna is basically irreligious and insensitive and frequently mocks at his beliefs. She provides no inspiration or appreciation of his creative impulses as a talented wood-carver. Will had found a mate for his body but not for his mind. Lawrence shows it as a failed marriage in which instead of companionship there is only rivalry and bitterness. A yawning emotional gulf arose between them that could not be bridged and each one began to lead the life of a recluse under the same roof.

Lawrence narrates the story of the third generation Bangwen Ursula in the final part of the novel in quest of fullness through her relationship with Anton Skrebensky, an army officer. Skrebensky is a highly career-oriented young man with his own aspirations in life which does not square with Ursula's values. They have a torrid love-affair where physical fusion fails to produce an emotional union. The absence of any vital and lasting connection between them leads to a mutual disenchantment between them. Skrebensky gets married to his Colonel's daughter and leaves England with his new wife to India. Ursula is left alone to pick up the pieces of her life. Lawrence thus points out the inadequacy of sexual passion by itself without shared goals and common interests to forge a lasting love-bond between a man and a woman.

Women in Love deals with the love-relationships of two sisters Ursula and Gudrun. Both Ursula and her lover Rupert Birkin had burnt their fingers in disastrous love affairs earlier, prior to meeting each other. Birkin's relationship with Hermione and Ursula's passion for Skrebensky had left them seared by its lack of wholeness. They emerged out of it sadder but wiser persons resolving to make wise choices in the future. Hermione's bloodless intellectuality repels Birkin whereas Ursula's wild sensual abandonment with Anton fails to produce any lasting pair-bond with him leading to the collapse of the relationship.

Birkin and Ursula are passionately in love with each other. They make painful adjustments to accept each other's differences. Finally Lawrence reveals Birkin and Ursula emerge through initial conflict, egotism and disenchantment towards a wonderful new life together as a married couple, achieving that elusive balance between maintaining one's individuality and staying committed to each other in a loving relationship.

The love-relationship between Gerald Crich and Gudrun on the other hand was doomed from the start. It was a romantic love based purely on physical attraction with no attempts made to check and comprehend the invisible qualities of character. Gerald is a dashing industrialist who can only charm women superficially without the capability to form a tender and lasting relationship with any woman. The romantic haze wears off quickly and the real self beneath the mask shows itself when Gerald and Gudrun are closeted together in an Austrian holiday resort. Gudrun finds him intolerably boorish, insanely jealous and capable of reckless violence even against her. She realizes that she had made a wrong choice in the most critical area of living. Gerald's brutal insensitivity, the abysmal lack of any aesthetic sense coupled with Gudrun. Even their making love degenerates into a corrosive deathly process of disintegration. Both frequently squabble with each other for dominance and this fierce clash of wills finally ends in revulsion and rejection of each other. Gudrun totally disillusioned, loses her interest in him and subsequently transfers all her affections on an artist named Loerke who makes her feel wanted and loved as a woman. Gerald is accidentally killed while skiing on the snowy mountains thus providentially relieving Gudrun from a relationship that had become a living hell.

In his most controversial novel *Lady Chatterley's Lover* Lawrence's primary aim was to treat human sexuality with great frankness so as to enable his readers to think rightly about sex as an integral element of their corporeal self. The central concern of Lawrence in all his major novels is the exploration of the dynamics of gender relationship with its inherent possibility of providing the basis for attainment of wholeness of being.

Lawrence's major purpose of writing *Lady Chatterley's Lover* is to help us find our way through sexuality's distortions into sexuality's wholeness. Lawrence's moral concern is transparent in the manner in which Lady Chatterley's affair with Oliver Mellors, the game-keeper is presented not as a glorification of adultery but as an exploration of valid alternatives to a confining and joyless marriage. Lawrence sought to demonstrate how an incomplete and loveless marriage in which there is a willful rejection of sexuality can lead to a painful and frustrating existence. Connie's major cause of unhappiness is her husband Clifford; who is paralysed below the waist and is incapable of satisfying his wife's normal and natural sexual desires. Unable to bear such extreme physical and emotional deprivation she unexpectedly stumbles on Mellors in her sprawling wooded estate and finds him a suitable lover who could awaken and transport her whole being into a higher plane of bliss and satisfaction. This intimate relationship provides her with the much needed sexual gratification, emotional attachment and spiritual healing after years of neglect by her callous husband. Lawrence does not encourage casual sex activity for in his view it reduces human beings to the purely animal and pant like dogs in heat. The novel concludes with Lawrence striking the right ethical note by making Connie and Mellors take responsibility for their actions and making their destiny by deciding to walk out of their unhappy marriages and seal their union with a life-long commitment and single-minded devotion to each other in marriage. The best remedy for the malady of an unrequited love is a new love for Lawrence. Thus Lawrence contrary to the opinions of his hostile critics never advocated immoral or unethical standards of behaviour in matters of marriage and sex.

Instead of misreading this novel and unfairly vilifying its author it is rewarding to be intellectually honest about the true value of this novel and its message. The core of *Lady Chatterley's Lover* is the analysis of the sexual aspect of love in a more open, frank and profound manner than any other novelist had done before. It is a magnificent work of verbal art which contains some of Lawrence's finest thinking on sex. Thus, it has been rehabilitated to its proper position as a work of literature by very discerning critics.

Thus in all his four major novels a clear strain of thought emerges from a close reading. Lawrence insistently and persistently maintained with great fervour that any wholeness of being or real living in a man-woman relationship was impossible to achieve on platonic terms. We humans are a combination of mind and body. Our bodies and souls are inseparably joined. It is possible to enjoy the fullness of life only by realizing sex as a privileged expression of love in a man-woman relationship in which there is a harmonious meeting of body and mind.

6

The Subversions of Postmodern Literature

The word "*postmodern*" is likely to have a polarizing effect around the discussion table. Postmodernism is a term that has been badly abused. Postmodernism is a capacious term within whose parameters a number of conflicting definitions and heterogeneous practices circulate. The term "*postmodernism*" is claimed to have been first used by the German Philosopher Rudolph Pennwitz to describe the nihilism of the twentieth-century culminating in Nietzsche.

Postmodernism is an elusive phenomenon that cannot be easily defined. It is a philosophy, culture an attitude, a social condition, a fad and a form of critique. It is a mood of our times. Some prominent intellectuals whose thinking has contributed to Postmodernism are Jean Francois Lyotard, Jaques Derrida, Michel Foucault and Jean Baudrillard.

Postmodernist literature is often characterized as Pop art which attempts to delink literature from the academic grooves and conventional modes of thinking. If modernism was a reaction against nineteenth-century romanticism and realism, postmodernism interrogates modernism as a vestige of the Arnoldian 'High culture' that presented the Humanist model of Man. The American critic Leslie Fielder considered postmodern literature as primarily a de-Eliotization movement in Art. He insisted that postmodernist literature must be free from the "finicky canons of the genteel tradition and the depressing pieties of the culture religion of Modernism".

Fielder stated that "the New-New Literature" must be aesthetic and poetic in form as well as substance but it must also be comical, irreverent and vulgar. The word vulgar is used here in the etymological meaning of the Latin word '*vulgus*' which means the common man. In that sense postmodernist literature has taken to re-vulgarization with a vengeance; it has striven to close the gap between art and popular entertainment, between elitist culture and popular culture by absorbing into its corpus, elements of contemporary popular culture, chiefly science fiction, pornography and the Western. By mid-fifties the modernist tendency in literature had run its course and postmodernism has since come to stay as literature, culture and theory. Postmodernist fiction is an international literary phenomenon which includes in its canon the Argentinian novelist Jorge Luis Borges, the Russian novelist Vladimir Nabokov, the chief French and English practitioners are Alain Robbe-Grillet and John Fowles. The well known American postmodern novelists are John Barth, Thomas Pyncheon and Norman Mailer.

Modernism believed in a kind of order. It believed in reconstructing the indispensable and relevant past. Modernists believed that the chaos that is human history can be given an order in Art. Eliot's past is a recreated and rediscovered past. Northrop Frye considered Myth as an ordering principle. Postmodernism denies linearity and stresses on discontinuity and seeks to destroy the past. To that extent postmodernism is schismatic. The distinguished literary critic Frank Kermode argues that "Total schism in arts is either communication or noise. If it is noise there is no more to say about it. If it is communication it is inescapably related to something older than itself". In the Modernist art of Eliot and Joyce, there was a clear distinction between art and entertainment. But, postmodernist art closes that gap between 'high' art and 'low' art through assimilation. The American postmodern novelist John Barth is of the view that all literary forms have been exhausted and writers of fiction need not concern themselves with reality at all as it is not an objective entity any more. From the "literature of exhaustion" one had to construct anew a "literature of replenishment". Writing fiction therefore is a conflict between the experiencing self and the formulating self. Some novelists like Norman Mailer and Truman Capote have experimented with different modes of fiction writing. They find contemporary reality so horrifying that they resort to fabulation and fantasy or just report life events in the form of non-fiction novel or reportage.

Fabulation is making up a story to cover the facts one does know or cannot accept. The novelist keeps a few facts and spins a new story round them. Truman Capote's In Cold Blood is not fiction but a report of a triple murder which occurred in Kansas city. Norman Mailer's Arunies of the Night is about his participation in the anti-Vietnam war march to the Pentagon. It is much like a news report or autobiography. The critic Robert Scholes describes fabulation as a postmodern narrative mode. The realistic and mimetic mode of fiction writing was rejected as reality was not monolithic and solidified. Postmodern writers employed fabulation as they challenged or repudiated all our erstwhile notions of tradition, truth and history. Fabulation is the result of a 're-cognition' of the amorphousness of reality.

Metafiction

Metafiction is the theory and practice of self-conscious fiction. These are "problematic novels" as fiction writing becomes the subject of the novel. Metafiction is fiction talking about fiction. It is fiction conscious of itself with a self-reflexivity that is concerned with the process of fiction making. Nineteenth-century fiction writers emphasized on social/ external realism. The French philosopher Paul Ricoeur in his essay *Mimesis and Representation* discusses Aristotle's account of mimesis and the poststructuralist critique of it. The conception of '*mimesis*' is central to fiction writing even today. Mimesis for Ricoeur is not a limited correspondence theory as Aristotle goes beyond Plato's conception of art as '*diegesis*' or talking of real actions and human passions. Aristotle combines mimesis with '*poesis*' thus stating that art is not a weakened copy of the world twice removed from the true Forms. It is not simply copying the already given but a creative production, a 're-*presentation*' and not imitation which brings about an augmentation of meaning by the writer's constructive activity.

The postmodern fiction writer is therefore engaged in 'an act of construction'. He is a maker of plots. Mimesis for him is more than imitation. It is a complex concept with many meanings like impersonation, portrayal, suggestion, expression, etc. It is an awareness of the complex interaction between language and the world. All reality is linguistically mediated and knowledge of the world cannot be immediate but must be filtered through some kind of conceptual frame of reference. We are intensely made aware of "the language consciousness" in the production of fiction.

Language is the tool that enables the fiction writer to create a quasi-real fictional world that allows readers to make sense of it by seeing how it connects with the world in which they live. 'Reality' in a metafictional narrative does not mean what is "already there and available". It is not copying a pre-existent world but effectively refiguring it by configuring it in order to transfigure our understanding of it. Metafictional works are not merely about the world but they produce in fictional form our own linguistic medication of it.

One of the best examples of metafiction is John Fowles' *The French Lieutenant's Woman.* There are many metafictional interventions in the novel which are typical features of post-war fiction. Fowles aims "to create a world that is". The love triangle between Charles and the two women he loves is depicted in terms of an existential conflict. Charles is irresistibly drawn towards the "sensual" Sarah with a "well-modeled and completely feminine face". He is bewitched by a 'fused rare power' emanating from her. She is described as a woman with "an unforgettable face" and with a stare "like a rifle or a lance". The contrast between the qualities of the two women shows Ernestina in a harsh light.

Sarah Woodruff is portrayed as being consistently authentic with a keen sense of perception. She is able to see people "as they were and not as they tried to seem". But in the twentieth chapter of the novel Fowles concludes with a question; "Who is Sarah? Out of what shadows does she come?"

Andrzy Gasiorek in his critical commentary on the novel observes that "Fowles' breaking of the fictional frame announces that his mock-Victorian novel is something of a *pastiche*. He thus suggests that it portrays nineteenth-century life in order to examine not only Victorian society as mediated by its literature but also the nature of fictional representation as it appears from a post-war perspective".

No answer is provided in chapter thirteen for the question raised at the end of the previous chapter. Only a metafictional enquiry on the problems of writing novels in the postmodern age is presented. A double-focus is offered here as in all metafictional texts. There is a simultaneous reference to a reality outside the text and to the process by which that reality is fictionally constructed. The core problem of representation is part of the subject matter of metafiction.

Historiographic Metafiction

Postmodernist fiction originated from the discourses of poststructuralism, postcolonialism, etc. It bred a New Fiction which created a revolution in the technique of writing fiction. There is no compulsion to impose order like Eliot's attempts to use myth to impose order on chaos. Distinctions between what is readable and unreadable collapse. Nabakov's *Lolita* which was earlier thought of as unabashed pornography is now considered a symbol of creativity. Thus intellectual culture is being redefined.

The New Fiction is influenced by the developments in Quantum Physics. We are in a relativist universe surrounded by a chaotic mass of energy. Nothing is stable and everything is in a state of uncertainty. The literary text is no more an objective entity to be analyzed as the new critics believed but it is a fluid nexus of possibilities to be experienced.

In metafiction, sur-fiction and narcissistic fiction, parody and irony are the staple techniques employed. The new fiction intends to subvert linguistic stability, historic objectivity and literary conventions and reorient the reader to new ways of thinking.

Historiographic metafiction is a play with history that problematizes history by pretending to be historical. History becomes a linguistic construct. It is itself a fiction. There is no observed which is not changed by the observer. The observer constantly changes the observed. Metafiction is self-reflexive art. It is art which is conscious of itself. Historiographic metafiction is a parody of history. It enables the reader to seriously understand history by a critical revisiting of history. It is a movement from the mimetic concept to the semiotic concept. There is no definite, objective social reality that can be recovered from the past but everything is interpreted through linguistic constructions. History is similar to the narratives of fiction as it is highly subjective and never neutral. It is a rhetorical account with no stable point of reference.

Salman Rushdie's *Midnight's Children*

Salman Rushdie is one of the greatest contemporary world writers. Reading his novels which are written in a postmodernist epic scale is an intellectual treat. Among contemporary writers Rushdie's fiction with its exuberant magic realism and postmodern fragmentations has the excitement of novelty. His influence on the literary world has been so profound as fictional works now belong to a pre-Rushdie or post-Rushdie ambience. His postcolonial mindscape and the hybridity of his background fits Homi Bhaba idea of "indeterminacy of diasporic identity". The paradox of his being "too English for the Indian" and "too Indian for the English" explains his rootless and disaffiliated status.

In *Midnight's Children*, Rushdie creates multiple realities of the world and then tries to navigate through them in his quest for new values. Each version of reality finds its own space without obliterating others. Magic realism is associated with Latin American literature. It is not a mode of narration but a mode of cognition. It is a way of life that mediates life through myths and legends and is contrary to scientific and rational thought.

In *Midnight's Children* you see reality constructed through fiction. There is no attempt made to bridge the gap between fiction and reality. History is seen as a play with reality and fiction as a play with history. Certain facts of history are deliberately distorted to prove this point. In a historical novel, the writer deals with certain events, facts, characters, etc. In historiographic metafiction certain unique events are parodied to realize unique dimensions hitherto unrealized in history. History is shown to be contradictory with so many versions. An event is narrated followed by the reactions to it. Which is the truth? Each one becomes a fabricator of the truth.

Rushdie's *Midnight's Children* is a great achievement of a fresh postmodern mode of writing. It is a major artistic production that reveals Rushdie not as a negative postmodernist but as someone who sees the disappearance of Truth as an opportunity, a chance to change the world for the better. The novel has two threads of narrative, as the protagonist Salim Sinai tells us. "The thread that leads to the ghettoes of the magicians and the thread that tells the story of Nadir, the rhymeless, verbless poet".

The novel proclaimed to contain one thousand and one stories of as many midnight's children born at the midnight hour of 15th August, 1947. Now these one thousand and one adventures are narrated by Salim Sinai, the protagonist who is man, de-man, and the demi-god Sherherazade. Drawing heavily on the magical realism of Gunter Grass's *The Tin Drum*, Rushdie makes the protagonist, Salim Sinai, float above history and time constantly decentering the outside as well as the inside of the nation. Rushdie, in order to demythologize the myths, twists them out of their shape. Salim's association with Padma and Parvati the witch, and reference to Dr. Aziz, Polly Umrigar, uncle Hanif Nassem Aziz, Mrs. Indira Gandhi, Vaijayanti Mala, and a host of other fictional and real life characters present the reader with a phantasmagoria of the fragmentary realities. The whole narrative covers a time-span of about sixty-two years from 1915 to 1977. Various fictional devices are used by Rushdie to recreate the history of the subcontinent.

Rushdie's narrative strategy in *Midnight's Children* is to juxtapose the personal and the public, the biographical and historical—a technique to be found throughout his novels. So the theme of 'the birth of a nation' has a parallel theme of the birth of the main protagonist Salim Sinai, who is also the first person narrator in the novel. Salim Sinai represents 1,001 more magical children supposed to have taken birth during the same hour, after the declaration of independence by Jawaharlal Nehru at the midnight hour of 15 August, 1947. Salim tells his story to Padma, a fellow worker in a pickle factory. The novel encapsulates the experiences of three generations of the Sinai family, living in Srinagar, Amritsar, Agra and Bombay and finally migrating to Karachi. Salim has the power of entering the minds and lives of others, and understanding their thoughts and feelings. He is, in fact an Anglo-Indian, the child of Vanita and Mr. Wee Willie Winkle. This is revealed by the Ayah, Mary Pereira, at Dr. Narlikar's home; Salim grows up in Bombay. He remembers his childhood and narrates that his grandfather Adam Aziz came from the paradise of Kashmir. He describes his grandfather's experience with Naseem whom he finally marries. Then, the family shifts from Kashmir to Agra where Adam's other daughters are married.

Salim remembers his childhood in Bombay. He recounts various political happenings in Bombay. Adam Aziz's story is related to the Indian struggle for Independence, from Jallianwala Bagh to the partition of India and is connected with every turn in the march of India's history.

In the second part of the novel, we are told about the growth and progress of Salim Sinai, the illegitimate child. Rushdie narrates the political, economical and social history of India. He takes us to Karachi, provides a few glimpses of Pakistan's political history and the Indo-Pak War is also described. Salim's parentage is revealed there.

Part three of the novel describes the onward march of Indian democracy. We have glimpses of various important political personalities and events in the novel. Leaders from Nehru to Mrs. Indira Gandhi are discussed. The creation of Bangladesh after the war, India's first nuclear explosion in

1974 and the declaration of Emergency in 1977 are mentioned. Salim marries Parvati and lives in one of Delhi's slums. He later moves with his wife and son to Bombay and takes up a job in a pickle factory. On completion of reading the novel, it is difficult to retain the whole story in mind as it is full of fissures and splinters of "contexts and inter-texts".

Historiographic metafiction tries to juxtapose history with fiction so that we are made conscious of the fiction of history and the reality of fiction. For the postmodernist novelist both reality and fiction are human constructs. Nothing is unchangeable or eternally given. There are no fixed boundaries between art and life, fiction and history. In historiographic metafiction, the boundaries of history and fiction are nullified. There is a blending of genres and autobiography comes in. The opposites are not reconciled. History and fiction confront each other with no resolution in sight.

Historiography or the writing of history is supposed to deal with facts and reality in a linear movement. Fiction usually deals not with facts but with imagination whereas history is concerned with what happened. Postmodern philosophers argue that history itself is a fiction and there is no objective history. There are only histories not a single history. There is no definite past and the past becomes the present in the flow of consciousness. All history is an interpretation. We are all participants in it.

In the *Midnight's Children,* Rushdie parodies history-making. Salim the central character is born on 15th August, 1947 exactly at 12 a.m. midnight. The two hands of the clock are folded in a greeting to welcome both him and the birth of a new nation on its Independence Day. Thus the history of a nation is interlinked with the history of an individual. The noise and shouts of joy during the celebrations of Independence Day outside the room are linked to his birth making both insignificant. Salim says of his birth. "I had been mysteriously handcuffed to history". History is thus reduced to a comedy.

There is a continuous crossing over of the individual into history and history into the individual. It is a non-linear narrative with a constant shuffling of time order. There is an illusion of continuity. The splitting of India is seen in the splitting of the individual. Salim claims, "I am the bomb in Bombay". A new perspective of history as a conscious fictional construct is provided. The reader gets lost in the labyrinthine maze of history. Reality masquerades as fiction. Stable identities are questioned and multiplicity in identity is celebrated. Thus in *Midnight's Children,* you have an episodic structure. There is an absence of a linear plat. The swoops and spirals, the digressions and reiterations are all characteristics of oral narrative.

The rambling text, the recollections of history and the intertwining of a smaller unit, the family with a larger unit, the community, shows Rushdie's technical brilliance. The reference to Indira Gandhi as a "Black widow" locates her as a historical figure and a monstrous fantasy. Different narrative techniques like mythic, filmic and magic realism are used. Thus richness of technique results in exciting multiple interpretative possibilities.

Magic realism which is the skillful combination of realism with a dream-like fantasy as in the case of the silver spittoon helps to crystallize Salim's sense of his personal history. The flying spittoon induces a state of amnesia in him. He loses his memory and consequently his identity. For memory is the chain which connects the postcolonial subject to his disrupted history. At the end, only fragments of history can be recollected. There is a riveting quality about the novel inspite of the deliberate anarchy created by the writer. The following passage is a striking example of the resolutely singular and the disconcertingly plural.

There are so many stories to tell, too many, such an excess of intertwined lives, events, miracles, places, rumours, so dense a co-mingling of the improbable and mundane. I have been a swallower of lives; and to know me, just the one of me, you will have to swallow the lot as well. Consumed multitudes are jostling and shoving inside me... (*MC*, p. 4).

Timothy Brennon observes that "Rushdie's novels are so uncompromisingly metafictional; they are novels about Third-world novels". There is a synthesis of fact and fiction. Fantasy becomes realistic and reality fantastic. Realism gets subverted by his genius strokes.

Julian Barnes as a Postmodernist Novelist

Julian Barnes is a novelist frequently associated with postmodernism—a literature marked by subversion techniques, deconstruction, fabulation and metafiction in preference to mimesis.

Though Barnes like Salmon Rushdie departs from realism in significant ways, he never rejects realist aesthetics and epistemology in their entirety. Their explorations of identity, gender, history and politics do not uphold postmodernism's more extreme pronouncements. In their novels one can see an exploration of the problem of history from a new set of perspectives.

Julian Barnes three highly acclaimed novels are *Flaubert's Parrot* published in 1985, *Staring at the Sun* (1986) and *A History of the World in 10½ Chapters* published in 1989. These novels were chosen by the poet Philip Larkin among his Books of the Year. Though *Flaubert's Parrot* was shortlisted for the Booker prize, it was according to a reviewer undoubtedly much too good to win the Booker prize. In 1986, after the publication of *Staring at the Sun,* Julian Barnes received the E.M. Forster Award from the American Academy of Arts and Letters. In both *Flaubert's Parrot and A History of the World in 10½ chapters,* Barnes explores the problem of history from a postmodern perspective.

Flaubert's Parrot focuses on the historian's inability to provide a reliable account of the past. The novel's narrator Geoffrey Braithwaite is unable to reconstruct the great French novelist's biography from a pile of fragments and details about the life of the creator of Madame Bovary. The quest for Flaubert's parrot, a symbol of the French novelist's authentic voice is doomed from the start:

> It isn't so different, the way we wander through the past. Lost, disordered, fearful, we follow what signs there remain; we read the street names, but cannot be

> confident where we are. All around is wreckage. A clock still ticks. A Parrot's perch catches the eye. We look for the parrot. Where is the Parrot? We still hear its voice, but all we can see is a bare wooden perch. The bird has flown.

Thus *Flaubert's Parrot* is a metafictional text which challenges the historicity of history. Metafiction is fiction that alerts the reader to the process of fiction-making or writing fiction. It is the practice of self-conscious fiction—art talking about itself. *Flaubert's Parrot* provides three alternative chronologies of Flaubert's life, which emphasize different key events and with multiple-narrative models invites the reader to see that there are different ways of reading documents and therefore several versions of the past with no stable record. Therefore, there can be no reliable account of the past, what one can hear is the Parrot's faint squawk but not its full speech. The fragments that are present point to what is absent with only partial accounts as vast areas of the past remain unaccountable.

Flaubert's Parrot is also a novel about interpretation. It not only discusses the difficulties of its own interpretation but also informs the reader how to interpret it. The problem of interpretation is presented this way:

> You can define a net in one of two ways, depending on your point of view. Normally, you would say that it is a meshed instrument designed to catch fish. But you could, with no great injury to logic, reverse the image and define a net as a jocular lexicographer once did; he called it a collection of holes tied together with a string. You can do the same with a biography.

Barnes here brilliantly demonstrates the first major thesis in Gadamerian hermeneutics that understanding is always horizonal. Human beings approach reality or a given literary work always from a given perspective or horizon. Horizon is the angle of vision or standpoint of an individual. It is the situatedness or context bound character of interpretation. When the search for the authentic Flaubert becomes impossible, the novel testifies to the problems of understanding with Braithwaite's insightful remarks. "The past", Braithwaite claims, is a piglet "smeared with grease; those trying to catch it fall over it while trying and are made to look ridiculous in the process". The novel avoids closure by concluding with "an answer and not an answer...an ending and not an ending". Therefore, the truth of Flaubert remains as slippery and elusive as his never to be discovered parrot. The value of such experimental fiction is that even its playfulness has a serious purpose. It packs a more powerful punch than traditional realist fiction. The novels of writers such as Barnes and Rushdie try not only to make sense of history and to distinguish between true and false accounts of it but also to confront the moral and political consequences of failing to do so.

In *Midnight's Children* Salim's redemptive fabulation is similarly foiled. He says "But reality is nagging at me. Love does not conquer all, except in the Bombay talkies".

And in Julian Barnes' *A History of the World in 10½ Chapters,* the novel confronts head-on with the difficulties of writing history. It is a novel full of fables and allegories, it asks whether history itself is a fiction or a form of fabulation. Fabulation is making up a story to cover the facts you don't know or can't accept. You keep a few true facts and spin a new story round them. History is just what historians tell us.

Fabulation is later explicitly linked to the writing of history in a chapter that deals with the problem of history. Barnes writes:

> History isn't what happened. History is just what historians tell us...one good story leads to another. First it was kings and archbishops with some offstage divine tinkering, then it was the march of ideas and movements of masses, then little local events...and we, the readers of history, the sufferers from history, we scan the pattern for hopeful conclusions, for the way ahead.

In the chapter "*The Survivor*", the protagonist rejects this account saying, "we've got to look at things how they are; we can't rely on fabulation anymore. It's the only way we'll survive". The conclusion Barnes arrives at is pragmatic.

Although the "God-eyed version" of truth is described as 'a charming impossible fake, the reader is urged to believe that 43% objective truth is better than 41% because the alternative is "beguiling relativity in which we value one liar's version as much as another liar's".

Thus, for Barnes political quietism is a surrender to fabulation but by refusing to differentiate the less accurate from the more accurate to submit to those who falsify history. He makes an impassioned plea for historical veracity and does not accede to postmodern skepticism. Rushdie in his *Imaginary Homelands: Essays on Criticism* asserts that "Literature must engage in debate over the nature of reality, for what is being disputed is nothing less than what is the case, what is truth and untruth". We cannot be forever playing multiple language games without taking a stance and to be constantly be engaged in an epistemological conflict by denying the possibility of reliable knowledge is cynical and even criminal on matters of life and death. Barnes too admits that objectivity may always elude us, but it is an "unattainable goal for which one must struggle in spite of the impossibility of success!".

Writers like Barnes have disproved the prophets of doom who had predicted 'the death of the novel' in the age of the Internet and cyberspace. In the hands of writers like Barnes and Rushdie the novel has reinvented itself. It has successfully met the onslaught of the mass media and not only survives but thrives. For the American postmodern philosopher Richard Rorty, we have no time for grand theories and in a "post-philosophy" phase, where philosophy is no more than a language game, his preference when it comes to finding a source of ideas to guide our behaviour is for other subjects such as literature.

According to Hans Georg Gadamer in his book *Truth and Method*, the experience of art is a matter of ontological disclosure rather than a matter of subjective consciousness. Literature is a verbal art form, which is fully, realized only when it is read. The text presents a sort of self-criticism, self-examination, a sort of participation in the world of the text which has brought about in us a new self-understanding.

Another great philosopher Paul Ricoeur points out the power of a literary text to project a world as well as to serve as a mediation of self-understanding. "To understand is to understand oneself before the text", wrote Ricoeur.

Thus contemporary literature gives us an intense vision of life and an intellectual awareness we would have otherwise never obtained. "Literature is the self-consciousness of society in permanent revolution" said Sartre. Great poets and novelists offer us the reassurance that somebody else sees things as we do or has the same kind of experience that we have. They perform the valuable service of challenging us to wonder whether our vision of the world is the right one.

7

The Long Delights of Short Story

The short story is a literary genre. It is usually fictional narrative prose and is more concise than longer works of fiction such as novels. The origins of short stories have been traced back to the traditions of oral story-telling and anecdotes. An anecdote is an amusing story about a real person or events. It was from the oral story telling traditions that provided the source for Homer's *Iliad* and *odyssey* those great epics of ware. These were often told in the force of rhyming poetry, which inspired the Homeric epithets. Later the short story evolved from the realistic novel as a miniature form. Leo Tolstoy and Anton Chekhov were considered as the finest Russian short story writers.

Some of the best short story writers in English were Kate Chopin, O. Henry, Ernest Hemingway, Katherine Mansfield, Flannery O' Connor, James Thurber and Edgar Allan Poe. Anecdotes were a kind of parable a brief realistic narration with a single message. Short stories are also considered to have originated from fables which are folk tales with an explicitly expressed moral. According to the Greek historian Herodotus, fables were invented by a Greek slave named Aesop in the sixth century B.C.

Chaucer's *Canterbury* Tales and Boccacio's *Decameron* developed into written stories in the early fourteenth century Europe from the oral story telling tradition. In the seventeenth century a collection of fairy tales was published by Charles Perrault.

In America, modern short stories emerged as a distinct genre in the early nineteenth century. The growth of magazines and journals created a strong demand for short stories having a word length of upto 15,000 words. The print media made short stories immensely popular Washington Irving's *Rip Van Winkle* (1819) and Nathaniel Hawthorne's *Twice-Told Tales* (1842) are the best examples. Hemingway wrote a long short story (called the novella) *The Old Man and the Sea.* In the twenty-first century, short stories appear in a dematerialized form in online magazines and on blogs.

The two well known early theorists of the short story are Edgar Allan Poe and Brander Matthews.

Elements and Characteristics

Poe's *The Philosophy of Composition* and Matthews' *The Philosophy of the Short Story* discuss the elements and characteristics of a short story. A classic definition of a short story is that one should be able to read it at one sitting (Poe). First, short stories are less complex than novels. A short story usually focuses on a single incident, has a single plot, a single setting, a limited number of characters and covers a short period of time.

Short stories contain a dramatic structure:

1. *Exposition*: the introduction of setting, situation and main characters.
2. *Complication*: the event in the story that produces a conflict.
3. *Crisis*: the decisive moment for the protagonist and his or her commitment to a sense of action as the conflict becomes most intense and the readers interest is at the highest.
4. *Resolution:* the final stage of the plot when the conflict is resolved and a moral lesson is learnt.

Modern short stories, because of their compact size, usually have an abrupt beginning and the story may even begin in the middle of the action. They do not follow the traditional pattern strictly. Some short stories have a 'twist in the tale' or a turning point and the endings may be abrupt and open without any definite message.

Anton Chekhov's *The Bet*, Kate Chopin's *The Story of an Hour, The Killers* by Ernest Heming way and *The Gift of the Magi* by O. Henry are immortal short stories to be read by every lover of literature.

A Psychoanalytic Reading of Kate Chopin's *The Storm*

There are three major characters in this short story. Bobinot, a man of humble origins, Calixta, his wife and his son Bibi. Bobinot loves his wife exceedingly and cares for her. Once when he got caught in a storm by surprise, he thinks of his wife Calixta at home and buys a can of shrimp which she likes. He holds the can as a rare treasure to give to his wife. He is sad that he cannot offer her the protection she needs during the storm. Even a brief separation unnerves him.

Calixta, in the meantime while watching the rain unexpectedly encounters her former lover Alcee who rides up to her home to seek shelter from the unexpected storm. Chopin brings to the reader's attention the mysterious inner workings of the human mind. We suddenly see Calixta clinging passionately to Alcee and abandons herself to a torrid love-making session. Everything is described in a non-judgemental, matter of fact manner with no hint of sordidness. When Bobinot brings home the shrimp, which is a symbol of his deep love for Calixta, she behaves as if nothing has happened. The two storms, one outer, the rain and the other inner, the storm of passion have passed and no one has been hurt as life goes on once again.

Freud stated that it is the repressed desires stored in our unconscious that control our conscious life. In the story, Calixta's Id which determines sexual drives overpowers her Ego, the conscious mind and makes her yield to forbidden passion. The Superego or the censor becomes powerless in this instance to rein in the Id. Calixta's act could be interpreted as the reflection of repressed physical appetites in her unconscious life.

A Formalist Reading of O. Henry's *The Gift of the Magi*

This short story by the famous American writer O. Henry is an eternal favourite with all short story readers. Jim and Della are a very loving and caring married couple from a working-class background. They wished to demonstrate their deep love by giving some expensive gifts to each other for Christmas. Like the Magi, or three Wise Men who brought gifts to the infant Christ, Della sold her beautiful hair to buy a platinum chain for Jim's watch. Jim on his part sold his watch to buy a set of expensive combs with jewelled rims. When the moment to exchange gifts arrived there is an unforeseen twist-in-the tale. Though the gifts were no more useful to them, O. Henry declares that "Of all who give and receive gifts, such as they are the wisest.... They are the Magi".

It is the sacrifice involved in the act of giving which is priceless than the cost or utility of the gift given. O. Henry's craftsmanship as a short story writer becomes evident from the construction of a clever plot with an unexpected climax. The reader is kept guessing till the last moment what the outcome is to be. This story is enriched by a blend of humanity and tender compassion that deeply affects the reader at the end.

Conclusion

Though the short story is a recent phenomenon in the history of literature, story telling is as old as the human race. Short story is an immensely popular form of literature today but it is doubtful if it would displace the novel in the near future. The short story cannot be considered as an abridgement of the novel. For the novel works on a larger canvas and deals with the whole life of individuals, generations of a family or a whole society. The novel is an epic in prose whereas the short story is a lyric in prose. Brevity and concentration are the hallmarks of a short story.

Unlike in a novel, all superfluous details are excluded in a short story. Unity of impression and rapidity of movement are indispensable for the effectiveness of a short story. Maximum effect should be achieved with minimum words. Not more than two or three characters can be presented in a particular situation. The social background can only be hinted with a couple of stokes. Short stories are essentially "vehicles of single impressions". Readers have a preference for short stories as a compact form of fiction which could be read easily and quickly. This is the secret of the short story's success as the most popular form of literature in our fast paced Information Age world of the 21st century.

References

Adeyanju. T.K. (1978). *Teaching Literature and Human Values in ESL*. ELT Journal 32/21.

Altick, Richard (1975). *The Art of Literary Research*. New York: Norton.

Ashcroft, Bill and Tiffin, Helen (eds.) (1989). *The Empire Writes Back: Theory and Practice in Post-colonial Literatures*. London: Routledge.

Booth, Wayne (1961). *The Rhetoric of Fiction*. Chicago: University of Chicago Press.

Bradley, A.C. (1914). *Shakespearean Tragedy*. London: Macmillan.

Ciardi, John (1975). *How does a Poem Mean*? Boston: Houghton Mifflin.

Cummings, M. and Simmons, R. (1983). *The Language of Literature*. London: Pergamon Press.

Dillon, G. (1978). *Language Processing and the Reading of Literature*. Indiana: Indiana University Press.

Duff, A. and Maley (1990). *Literature*. Scotland: OUP.

Eagleton, Terry (2003). *After Theory*. New York: Basic Books.

Fowler, R. (1981). *Literature as Social Discourse*. London: Batsford.

Gower, R. and Pearson, M. (1986). *Reading Literature*. London: Longman.

Granville Barker, H. (1946). *Prefaces to Shakespeare*. Princeton: Princeton University Press.

Huyseen, Andreus (1988). *After the Great Divide: Modernism, Mass Culture, Postmodernism*. Bloomington: Indiana University Press.

Jones, R.T. (1986). *Studying Poetry*. London: Edward Arnold.

Jameson, Frederic (1991). *Postmodernism, or the Cultural Logic of Late Capitalism*. Durham: Duke University Press.

Leavis, F.R. (1969). *English Literature in Our Time and the University*. London: Chatto and Windus.

Laurenson, Diana and Swingwood, Alan (1972). *The Sociology of Literature*. London: Paladin.

Leech, G.N. and Short, M.H. (1981). *Style in Fiction*. London: Longman.

Mailloux, Steven (1984). *Interpretive conventions: The Reader in the Study of American Fiction*. Ithaca: Cornell University Press.

Moody, H.L.B. (1971). *The Teaching of Literature*. London: Longman.

Morrison, Toni (1993). *Playing in the Dark: Whiteness and Literary Imagination*. New York: Vintage Press.

Page, N. (ed.) (1984). *The Language of Literature*. London: Macmillan.

Pechter, Edward (1987). *The New Historicism and its Discontents: Politicizing Renaissance Drama*. PMLA 102 (May 1987): 292-303.

Radmay, Janice (1984). *Reading the Romance: Women, Patriarchy and Popular Literature*. Chapel Hill: University of North Carolina Press.

Richter, David (ed.) (1998). *The Critical Tradition*. New York: St. Martin's.

Said, Edward (1978). *Orientalism*. New York: Pantheon.

Short, M.H. (ed.) (1988). *Reading, Analyzing and Teaching Literature*. London: Longman.

Veeser, H. Aram (ed.) (1989). *The New Historicism*. New York: Routledge.

Walker, R. (1983). *Language for Literature*. London: Collins.

Widdowson, H.G. (1975). *Stylistics and the Teaching of Literature*. London: Longman.

White, Hayden (1973). *Metahistory: The Historical Imagination in Nineteenth Century Europe*. Baltimore: Johns Hopkins University Press.